Electrochemistry II

Electrochemical Techniques in Bioanalysis

C. E. Lunte and W. R. Heineman

Medical Applications of Electrochemical Sensors and Techniques

G. S. Calabrese and K. M. O'Connell

Photoelectrochemical Solar Energy Conversion

R. Memming

Mechanism of Reactions on Colloidal Microelectrodes and Size Quantization Effects

A. Henglein

THEORIA CUM PRAXI

AKADEMIE-VERLAG BERLIN

Topics in Current Chemistry · Vol. 143

Electrochemistry II

Electrochemistry II

With 79 Figures and 9 Tables

AKADEMIE-VERLAG BERLIN
1988

Die Originalausgabe erscheint im Springer-Verlag
Berlin · Heidelberg · New York · London · Paris · Tokyo
in der Schriftenreihe „Topics in Current Chemistry",
Volume 143

Vertriebsrechte für die sozialistischen Länder:
Akademie-Verlag Berlin

Vertriebsrechte für alle Staaten mit Ausnahme der sozialistischen Länder:
Springer-Verlag Berlin · Heidelberg · New York · London · Paris · Tokyo

ISBN 3-540-18226-8 Springer-Verlag Berlin · Heidelberg · New York
ISBN 0-387-18226-8 Springer-Verlag New York · Heidelberg · Berlin

ISBN 3-05-500502-3 Akademie-Verlag Berlin

Erschienen im Akademie-Verlag Berlin, DDR-1086 Berlin, Leipziger Straße 3—4
Lizenznummer: 202 · 100/505/88
Printed in the German Democratic Republic
Gesamtherstellung: VEB Druckerei „Thomas Müntzer", 5820 Bad Langensalza
LSV 1215
Bestellnummer: 763 833 5 (3072/143)
13800

Editorial Board

Preface to the Series on Electrochemistry

The scope of electrochemistry having broadened tremendously within the last ten years has become a remarkably diverse science. In the field of electroorganic synthesis, for example, selectivity has been improved by use of electrogenerated reagents, energy uptake lowered and space-time yields have been improved by using mediated reactions. In addition, electroorganic chemistry has been efficiently applied to the synthesis of key building blocks for complex molecules and has established its role as a new tool in organic synthesis. However electrochemistry has also found new and interesting applications in quite different fields of chemistry. Photoelectrochemistry, as one example, is not only valuable for transformations of organic molecules but also for the very important goal of energy conversion. More insight has been gained in the processes occurring on illuminated semiconductor electrodes and micro particles. Designing the composition of electrode surfaces can lead to the selective activation of electrodes. Electrochemical sensors and techniques present new opportunities for the analysis of biological compounds in medicine and biology. Research in the field of conducting polymers is very intensive because of interesting potential applications.

Therefore I am very happy that Springer-Verlag has decided to account for these important developments by introducing a series of volumes on new trends in electrochemistry within its series Topics in Current Chemistry. The volumes will cover the important trends in electrochemistry as outlined above in the following manner:

Electroorganic Synthesis by Indirect Electrochemical Methods;
New Applications of Electrochemical Techniques;
Recent Development in Electroorganic Synthesis.

The guest editor is very happy and thankful that well-known experts who are actively engaged in research in these fields have agreed to contribute to the volumes. It is hoped that this collection of reviews is not only valuable to investigators in the respective fields but also to many chemists who are not so familiar with electrochemistry.

Bonn, Mai 1987 Eberhard Steckhan

Preface to Volume II

Two volumes of the electrochemistry series in Topics in Current Chemistry are dedicated to new applications of electrochemical techniques. The contributions in Volume II cover two fields:

— Application of electrochemical methods as analytical tools for the detection as well as the concentration and activity determination of biologically active compounds in bioanalysis and medicine;
— Electrochemical processes for photoelectrochemical energy conversion using semiconductor electrodes or microparticles.

In the first two contributions electroanalytical techniques are described for application in bioanalysis and medicine. The increasing interest in this field is mainly due to the excellent selectivities and detection limits. In addition, the possibilities of miniaturization allow the development of in vivo analysis.

The following two papers deal mainly with problems in energy conversion, in particular, the transformation of irradiation energy into electrical or chemical energy. The present status and future possible developments of photoelectrochemical energy conversion is presented. In a second paper electrochemical developments are connected to colloidal chemistry and the application of colloidal particles as catalysts for electron transfer reactions and as photocatalysts are discussed.

These articles may show that electrochemical developments and studies not only influence a restricted area but are also important for a number of other fields.

Bonn, June 1987 Eberhard Steckhan

Table of Contents

Table of Contents of Volume 142

Electrochemical Techniques in Bioanalysis

Craig E. Lunte* and William R. Heineman

Department of Chemistry, University of Cincinnati, Cincinnati, Ohio 45221-0172, U.S.A.

Table of Contents

* Current Address: Department of Chemistry, The University of Kansas, Lawrence, KS 66045-0046, U.S.A.

Electroanalytical techniques are currently gaining popularity in the area of biochemical analysis. There are several reasons for this growth. Electrochemical techniques provide both excellent detection limits with wide dynamic range. These methods are typically readily amenable to miniaturization. This leads to being able to use very small sample volumes. In vivo analysis is also possible. Finally, electrochemical techniques are generally very selective and when not selective enough can easily be coupled to even more selective techniques such as liquid chromatography or immunoassay. In this chapter, the basic concepts underlying electroanalytical methods of bioanalysis will be described. How these techniques can be used in bioanalytical chemistry will be illustrated. In addition, the interrelationship of the various electroanalytical techniques will be discussed.

1 Introduction

Electroanalytical techniques are based on the measurement of electrical signals associated with electrochemical cells. Two basic categories of techniques exist — potentiometric and dynamic. Potentiometric techniques involve the measurement of a cell potential under equilibrium conditions in which no electrolysis is occurring at the electrodes. The potential of the cell is related to the concentration of analyte in the sample. By comparison, dynamic techniques are based on measurements made on a cell in which electrolysis is occurring, i.e. reduction or oxidation of the analyte at an electrode. Generally, a potential is being applied to the cell (excitation signal) and the resulting current (response signal) is proportional to the concentration of analyte in the sample. A wide variety of potentiometric and dynamic electroanalytical techniques have been developed. These techniques differ in cell configuration, excitation signal and response signal. This monograph deals with those techniques that have proved to be especially useful in bioanalysis.

Electroanalytical techniques have certain features that are advantageous for bioanalysis. Many of the techniques exhibit excellent detection limits coupled with a wide dynamic range. Methods are available for the determination of the relatively high concentrations found for species such as blood electrolytes (Na^+, K^+, Cl^-, HCO_3^-) as well as trace levels of species such as metabolites in blood and urine samples and heavy metals in crop samples. Measurements can generally be made on very small samples, typically in the microliter volume range. Coupling the good detection limits with the small volume capability enables determinations to be made on down to sub-picogram amounts of analyte. Some electroanalytical techniques exhibit excellent selectivity, which is especially important in the analysis of complicated samples of biological origin. By comparison to spectroscopic techniques such as UV-visible absorption and fluorescence, a sample generally contains fewer electrochemical than spectroscopic interferents. Electroanalytical techniques can be used in vivo. For example, miniature electrochemical sensors are used to measure pH and pO_2 with in-dwelling catheters. Electroanalysis is an inexpensive technique in comparison to most instrumental methods. Some electrochemical cells have been made sufficiently inexpensive to be disposable items — $ 1.25 each. The cost of ancillary instrumentation ranges from ca. $ 200 for a pH meter to $ 40,000 for a sophisticated, multi-technique instrument.

2 Potentiometry

Potentiometric methods are based on the measurement of the potential of an electrochemical cell consisting of two electrodes immersed in a solution. Since the cell potential is measured under the condition of zero current, usually with a pH/mV meter, potentiometry is an equilibrium method. One electrode, the indicator electrode, is chosen to respond to a particular species in solution whose activity or concentration is to be measured. The other electrode is a reference electrode whose half-cell potential is invariant.

The indicator electrode is of paramount importance in analytical potentiometry. This electrode should interact with the species of interest so that the potential of the electrochemical cell reflects the activity of this species in solution and not other compounds in the sample that might constitute an interference. The importance of having indicator electrodes that selectively respond to numerous species of analytical significance has stimulated the development of many types of indicator electrodes.

Historically, indicator electrodes have been metals which form a redox couple with the analyte, such as a Ag electrode for the determination of Ag^+, or a chemically inert metal which responds to the activity ratio of a soluble redox couple, such as a Pt electrode for Fe^{3+}/Fe^{2+}. Whereas simple indicator electrodes of this type perform well for the analysis of relatively pure samples, they are often subject to interferences when applied to complex samples such as those of biological origin.

2.1 Ion-Selective Electrodes

A significant development in the methodology of potentiometry that paved the way for its utility in bioanalysis was the discovery of the ion selective electrode (ISE). Conceptually, the ISE involves the measurement of a membrane potential. The response of the electrochemical cell is therefore based on an interaction between the membrane and the analyte that alters the potential across the membrane. The selectivity of the potential response to the analyte depends on the specificity of the membrane interaction for the analyte.

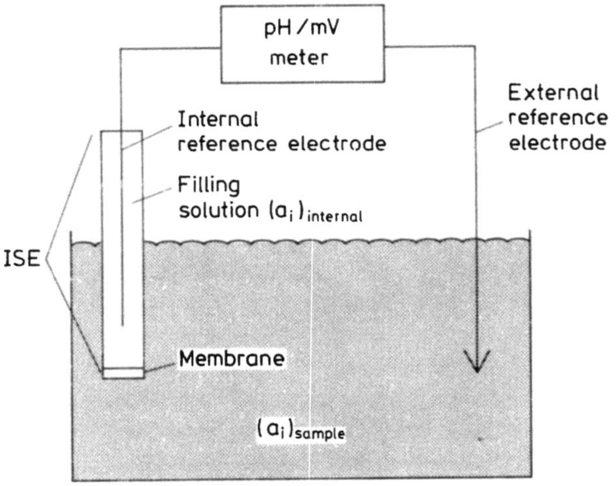

Fig. 1. Schematic diagram of ion-selective electrode (ISE)

A representative ISE is shown schematically in Fig. 1. The electrode consists of a membrane, an internal reference electrolyte of fixed activity, $(a_i)_{internal}$ and an internal reference electrode. The ISE is immersed in sample solution that contains analyte of some activity, $(a_i)_{sample}$ and into which an external reference electrode is also immersed. The potential measured by the pH/mV meter (E_{cell}) is equal to the difference in potential between the internal ($E_{ref,int}$) and external ($E_{ref,ext}$) reference electrodes, plus the membrane potential (E_{memb}), plus the liquid junction potential

(E_{lj}) that exists at the junction between the external reference electrode and the sample solution.

$$E_{cell} = E_{ref,\,int} - E_{ref,\,ext} + E_{memb} + E_{lj} \tag{1}$$

If the membrane is permselective for a particular ion (i), a potential develops across the membrane which depends on the ratio of activities of the ion on either side of the membrane as described by the Nernst equation

$$E_{memb} = \frac{RT}{zF} \ln \frac{(a_i)_{sample}}{(a_i)_{internal}} \tag{2}$$

where z is the charge on the ion, R is the gas constant (8.314 J mol^{-1} K^{-1}), T is temperature (K), and F is the Faraday (96,485 C mol^{-1}). Substitution of Eq. (2) into Eq. (1) gives an overall expression for the electrochemical cell that can be written as follows:

$$E_{cell} = E_{ref,\,int} - E_{ref,\,ext} + \frac{RT}{zF} \ln \frac{1}{(a_i)_{internal}} + E_{lj} + \frac{RT}{zF} \ln (a_i)_{sample} \tag{3}$$

The half-cell potentials of the two reference electrodes are constant; sample solution conditions can often be controlled so that E_{lj} is effectively constant; and the composition of the internal solution can be maintained so that $(a_i)_{internal}$ is fixed. Consequently Eq. (3) can be simplified to give

$$E_{cell} = K + \frac{0.0591}{z} \log (a_i)_{sample} \tag{4}$$

at 25 °C where K represents the constant terms in Eq. (3). This logarithmic relationship between cell potential and analyte activity is the basis of the ISE as an analytical device. A plot of E_{cell} vs. log a_i for a series of standard solutions should be linear over the working range of the electrode and have a slope of 0.0591/z for measurements made at 25 °C.

Since membranes respond to a certain degree to ions other than the analyte (i.e., interferents), a more general expression than Eq. (4) is

$$E_{cell} = K + \frac{0.0591}{z} \log (a_i + k_{ij} a_j^{z/x}) \tag{5}$$

where a_j is the activity of interferent ion, j, x is the charge of the interferent ion, and k_{ij} is the selectivity constant. Small values of k_{ij} are characteristic of electrodes with good selectivity for the analyte, *i*.

The development of successful ISEs has hinged on the search for membranes that exhibit both sensitivity and selectivity for the species of interest. Of the two desirable properties, selectivity is by far the more difficult to achieve. ISEs with selectivity for cations and anions have been developed with three basic types of membranes: liquid and polymer, solid state, and glass. Electrodes are commercially available for numerous ions including H^+, Li^+, Na^+, K^+, Ag^+, Ca^{2+}, Cu^{2+}, Pb^{2+}, Cd^{2+}, water hardness ($Ca^{2+} + Mg^{2+}$), F^-, Cl^-, Br^-, I^-, SCN^-, NO_3^-, ClO_4^-, and BF_4^-. Additionally, electrodes have been reported for numerous organic anions and cations [1,2]. ISEs have found wide application in biomedical and clinical laboratories throughout the world. This important application is discussed in Chapter X. ISEs also constitute the basis for gas sensing electrodes and many biosensors, which are discussed in the next two sections. A number of excellent monographs have been written on ISEs [3-12].

2.2 Gas-Sensing Electrodes

Gas-sensing electrodes consist of an ion-selective electrode in contact with a thin layer of aqueous electrolyte that is confined to the electrode surface by an outer membrane as shown schematically for a CO_2 electrode in Fig. 2. The outer membrane,

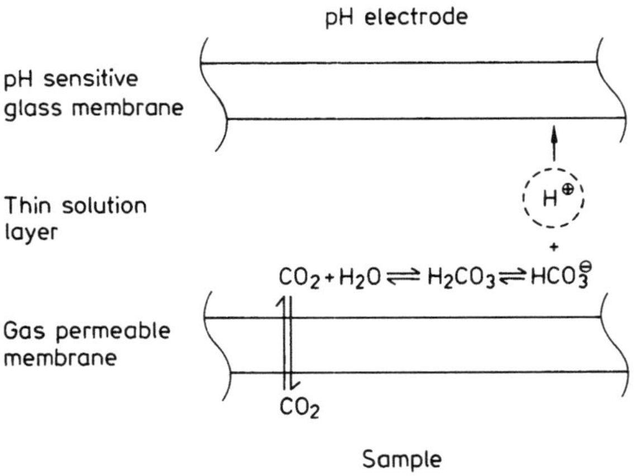

Fig. 2. Schematic representation of gas-sensing electrode for CO_2

which is chosen to be permeable to the gas, is silicone rubber for CO_2. Exposure of the electrode to a sample causes CO_2 gas to pass through the membrane. Dissolution of the CO_2 in the thin layer of electrolyte causes a change in pH due to a shift in the equilibrium position of the chemical reaction shown in Fig. 2. The change in pH sensed by the underlying pH electrode is in proportion to pCO_2 of the sample. Elec-

Table 1. Gas-sensing electrodes

Gas electrode	Internal solution equilibrium	Sensing
CO_2	$CO_2 + H_2O \rightleftharpoons HCO_3^- + H^+$	Glass, pH
NH_3	$NH_3 + H_2O \rightleftharpoons NH_4^+ + H^+$	Glass, pH
HCN	$HCN \rightleftharpoons H^+ + CN^-$	Ag_2S, pCN
HF	$HF \rightleftharpoons H^+ + F^-$	LaF_3, pF
H_2S	$H_2S \rightleftharpoons 2\,H^+ + S^{2-}$	Ag_2S, pS
SO_2	$SO_2 + H_2O \rightleftharpoons HSO_3^- + H^+$	Glass, pH

trodes for a variety of gases have been developed based on the internal solution equilibria and sensing electrodes shown in Table 1. One of the most important applications of the CO_2 electrode is the measurement of blood pCO_2 as discussed in Chapter X. Gas electrodes have been used for other bioanalytical applications such as the measurement of CO_2 in general assays of decarboxylating enzyme activities [13] and the measurement of NH_3 in meat tissue [14] and serum [15].

2.3 Biocatalytic Membrane Electrodes: Biosensors

Biocatalytic membrane electrodes have an ISE or a gas sensing electrode in contact with a thin layer of biocatalytic material, which can be an immobilized enzyme, bacterial particles or a tissue slice, as shown in Fig. 3 [16–19]. The biocatalyst converts substrate (the analyte) into product, which is measured by the electrode. Electrodes of this type are often referred to as "biosensors".

A representative example of a biocatalytic membrane electrode is an electrode for L-arginine [20]. The bacterium *streptococcus faecium* is immobilized on the gas permeable membrane of an ammonia electrode. Arginine deiminase in the bacterium catalyzes the following reaction

$$\text{L-arginine} + H_2O \xrightarrow[\text{deaminase}]{\text{arginine}} \text{citrulline} + NH_3 \qquad (6)$$

The electrode responds to the concentration of NH_3 at its surface which is proportional to the concentration of L-arginine in the sample. Nernstian plots give slopes of 40 to 45 mV with a linear range of 1×10^{-4} to 6.5×10^{-3} M. The electrode is highly selective for L-arginine until contaminants grow in the biocatalyst layer, which initiates response to glutamine and asparagine. The growth of such contaminants can be discouraged under certain storage conditions [21]. Numerous examples of other biocatalytic electrodes have been reported, some of which are listed in Table 2.

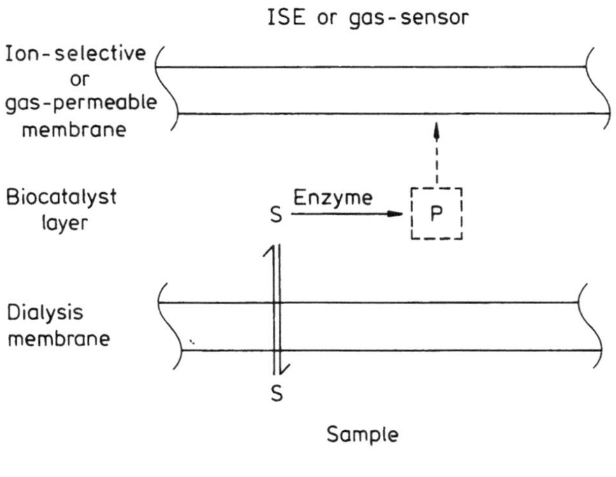

Fig. 3. Schematic diagram of biocatalytic electrode

Table 2. Biocatalytic membrane electrodes[a]

Biocatalyst category	Substrate	Biocatalyst	Detected substance	Ref.
Bacterial particle	L-arginine	Streptococcus faecium	NH_3	20)
		Streptococcus lactis	NH_3	43)
	L-aspartate	Bacterium cadaveris	NH_3	22)
	L-cysteine	Proteus morganii	H_2S	33)
	L-glutamine	Sarcina flava	NH_3	23, 86)
	L-glutamic acid	Escherichia coli	CO_2	31)
	L-histidine	Pseudomonas sp.	NH_3	27)
	L-tyrosine	Aeromonas phenologenes	NH_3	26)
	L-serine	Clostridium acidiurici	NH_3	28)
	NAD^+	Escherichia coli/NADase	NH_3	24)
	Nitrilotriacetate acid	Pseudomonas sp.	NH_3	25)
	Nitrate	Azotobacter vinelandii	NH_3	29)
	Uric acid	Pichia membranaefaciens	CO_2	30)
	Pyruvate	Streptococcus faecium	CO_2	32)
	Sugars	Bacteria from human dental plaque	H^+	34)
Tissue	Glucosamine 6-phosphate	Porcine kidney tissue	NH_3	35)
	Adenosine	Mouse small intestine mucosal cells	NH_3	37)
	Adenosine 5'-monophosphate	Rabbit muscle	NH_3	38)
	L-cysteine	Treated plant leaf		36)
	L-glutamate	Yellow squash	CO_2	39)
	Glutamine	Porcine kidney cortex	NH_3	40, 41, 86)
	Guanine	Rabbit liver	NH_3	42)
Enzyme	Urea	Urease	NH_4^+	44–47, 48, 101)
			NH_3	49–52, 91–93, 110, 111)
			CO_2	53, 54)
			H^+	55, 90, 108, 109)
	Glucose	Glucose oxidase and peroxidase	I^-	56, 96)
		Glucose oxidase	H^+	55)
		Glucose oxidase	F^-	97)
	Amygdalin	β-glucosidase	CN^-	57, 58, 59)
	Pencillin	Penicillinase	H^+	55, 60, 61)
	L-amino acids	L-amino acid oxidase	NH_4^+	62, 63)
	D-amino acids	D-amino acid oxidase	NH_4^+	64)
	L-alanine	NAD, alanine dehydrogenase, and lactate dehydrogenase	NH_3	102)
	Arginine	Arginase and	NH_3	84)
		urease	NH_3	104)
	Asparagine	Asparaginase	NH_4^+	64)
			NH_3	70)
	Glutamine	Glutaminase	NH_4^+	67)
			NH_3	86)

Table 2. (continued)

Biocatalyst category	Substrate	Biocatalyst	Detected substance	Ref.
	Glutamate	Glutamate dehydrogenase and lactate dehydrogenase	NH_4^+	72)
		Glutamate decarboxylase	CO_2	73, 98)
	L-histidine	Histidine ammonia-lyase	NH_3	88)
	L-lysine	L-lysine decarboxylase	CO_2	65, 66, 71, 106)
	Methionine	Methioninelyase	NH_3	112)
	N-Acetyl-L =ethionine	Acylase and L-amino acid oxidase	NH_3	103)
	L-phenylalanine	L-amino acid oxidase	NH_4^+	68)
		L-amino acid oxidase and horseradish peroxidase	I^-	68)
		Phenylalanine ammonia lyase	NH_3	69)
		Phenylalanine decarboxylase	CO_2	65, 66)
	L-tyrosine	Tyrosine decarboxylase	CO_2	53, 65, 66, 107)
	5'-adenosine monophosphate (AMP)	AMP deaminase	NH_3	74)
	Cyclic AMP	Phosphodiesterase and AMP deaminase	NH_3	75)
	Uric acid	Uricase	CO_2	76, 94)
	Creatinine	Creatinine deaminase	NH_3	77, 85, 105)
	Acetylcholine	Acetylcholinesterase	H^+	78, 79)
	D-gluconate	Gluconate kinase and 6-phospho-D-gluconate	CO_2	80)
	Lactate	Lactate dehydrogenase	ferri-ferrocyanide (redox)	81)
	Acetaldehyde	Aldehyde dehydrogenase	H^+	82)
	Adenosine	Adenosine deaminase	NH_3	83)
	Guanine	Guanase	NH_3	87)
	Oxalate	Oxalate decarboxylase	CO_2	89)
	Flavin adenine dinucleotide	Alkaline phosphatase and adenosine deaminase	NH_3	95)
	Methotreaxate	Dihydrofolate reductase and 6-phosphogluconic dehydrogenase	CO_2	99)
	Salicylate	Salicylate hydroxylase	CO_2	100)

[a] Compiled from 1, 2, 16, 17, 19

Biocatalytic membrane electrodes significantly expand the scope of direct potentiometry by enabling biosensors that respond to a whole host of organic substrates to be made. The selectivity of these sensors is a combination of the selectivity of the biocatalyst for the substrate and the ISE for other constituents in the sample that might reach the ISE surface membrane. Thus, the selectivity with respect to other organic constituents in the sample is determined by biocatalyst specificity whereas the selectivity with respect to inorganic ionic or gaseous constituents in the sample is determined by the underlying ISE or gas sensor. Electrodes such as that for L-arginine exhibit excellent selectivity due to the high specificity of the enzyme arginine deiminase for

L-arginine and the excellent selectivity of the ammonia electrode for NH_3. The working range of biosensors of this type is typically only two to three orders of magnitude with a detection limit of 10^{-5} to 10^{-4} M.

Cell-based biosensors have comparable response characteristics to enzyme electrodes, but offer several advantageous features [16]. a) Strains of bacteria are generally less expensive than isolated enzymes. A specific sterile bacterial culture is not needed to prepare an electrode. For example, human dental plaque suffices to prepare electrodes for D(+) glucose, D(+) mannose and D(—) fructose [34]. b) Enzyme activity is often enhanced in bacterial cells and remains longer due to the optimal environment. Consequently, lifetimes of bacterial electrodes average around 20 days compared to 14 days for enzyme electrodes. Cells can be regrown if the catalytic activity is lost. c) Bacteria can readily effect complex reaction sequences requiring cofactors. For example, an electrode for nitrilotriacetic acid (NTA) is practical only because *pseudomonas* bacterial cells contain all of the enzymes and cofactors necessary to execute the following sequence of reactions on which the electrode response is based [25]:

$$\begin{aligned} \text{NTA} + \text{NADH} + O_2 &\xrightarrow[\text{monooxygenase}]{\text{NTA}} (\alpha\text{-hydroxy-NTA}) && (7) \\ &\xrightarrow{\hspace{2em}} \text{iminodiacetate} + \text{glyoxylate} \\ \text{iminodiacetate} &\xrightarrow{\hspace{2em}} \text{glycine} + \text{glyoxylate} \\ \text{glycine} &\xrightarrow[\text{decarboxylate}]{\text{glycine}} \text{serine} + CO_2 \\ \text{serine} &\xrightarrow[\text{deaminase}]{\text{serine}} \text{pyruvate} + NH_3 \end{aligned}$$

The underlying gas electrode for ammonia, responds to NH_3 formed in the last reaction, which is proportional to the concentration of NTA in the sample.

The concept of a biocatalytic membrane electrode has been extended to the use of a tissue slice as the catalytic layer [18]. An example of this approach is an electrode for AMP which consists of a slice of rabbit muscle adjacent to an ammonia gas electrode [38]. NH_3 is produced by enzymatic action of rabbit muscle constituents on AMP The electrode exhibits a linear range of 1.4×10^{-4} to 1.0×10^{-2} M with a response time varying from 2.5 to 8.5 min, depending on the concentration. Electrode lifetime is about 28 days when stored between use in buffer with sodium azide to prevent bacterial growth. Excellent selectivity enables AMP to be determined in serum.

Cell and tissue based electrodes are generally less selective than enzyme based electrodes since the many constituents of cells result in response to many possible substrates. This problem may be solved by the use of specific inhibitors to block interfering reactions. These electrodes also suffer from long recovery times (3–4 hours) after a measurement has been made. However, such electrodes may be sufficiently cheap to be disposable. Most biocatalytic membrane electrodes are susceptible to irreversible loss of activity when exposed to samples containing denaturing or toxic agents. The difficulties associated with electrode longevity, both during use and storage, have stymied the commercial development of biocatalytic membrane electrodes.

2.4 Ion Sensitive Field Effect Transistors (ISFET)

The ISFET is an electrochemical sensor based on a modification of the metal oxide semiconductor field effect transistor (MOSFET). The metal gate of the MOSFET is replaced by a reference electrode and the gate insulator is exposed to the analyte solution or is coated with an ion-selective membrane [113, 114] as illustrated in Fig. 4 [115].

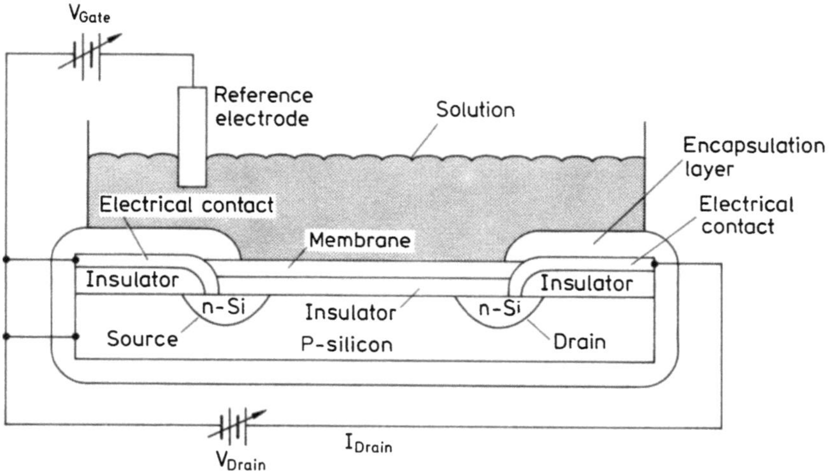

Fig. 4. Schematic diagrams of an ISFET

The rest of the device is protected by a suitable encapsulant. Insulators such as SiO_2, Al_2O_3 and Ta_2O_5 have surface hydroxyl groups that act as sites for chemical reactions when exposed to an electrolyte solution. The alteration in surface charge resulting from protonation/deprotonation of these surface sites affects the surface potential and results in a relation between the drain current of the ISFET and solution pH. For example, an Al_2O_3-ISFET exhibits a linear response between amplifier voltage and pH over a range of *ca.* 2.5 to 10.5 with a slope of 53 mV/pH [113]. Ion-sensitive membranes coated on the insulator surface have been used to achieve selective response to other cations and anions (e.g., H^+, K^+, Ca^{2+}, Cl^-, I^- and CN^-) [116]. Coatings of immobilized enzymes have been used in a manner that is analogous to their use with the ISE. For example, an ISFET coated with penicillinase immobilized in a membrane responds to penicillin due to the local pH change resulting from the enzyme-catalyzed formation of penicilloic acid [117]. Enzymatically coupled pH ISFETS have also been used for the measurement of urea [118], acetylcholine [118], and glucose [119]. The main interest in ISFETs stems from the ease with which inexpensive microelectrodes can be produced. However, the commercial success of these devices has been hindered by problems with encapsulation and stability.

2.5 Miniature Electrodes and in vivo Measurements

A significant characteristic of the ISE is the feasibility of fabricating miniature sensors for measurements in samples of very small volume or in vivo [120–123]. Microelectrodes with tip diameters ranging from 0.1 to 200 μm have been devised using glass electrodes for H^+ [124], K^+ and Na^+ [125], solid state electrodes for Ag^+, Cu^{2+}, I^-, Cl^-, Br^- and S^{2-} [126] and liquid membrane electrodes for K^+, Cl^-, Na^+ [127], Mg^{2+} [128],

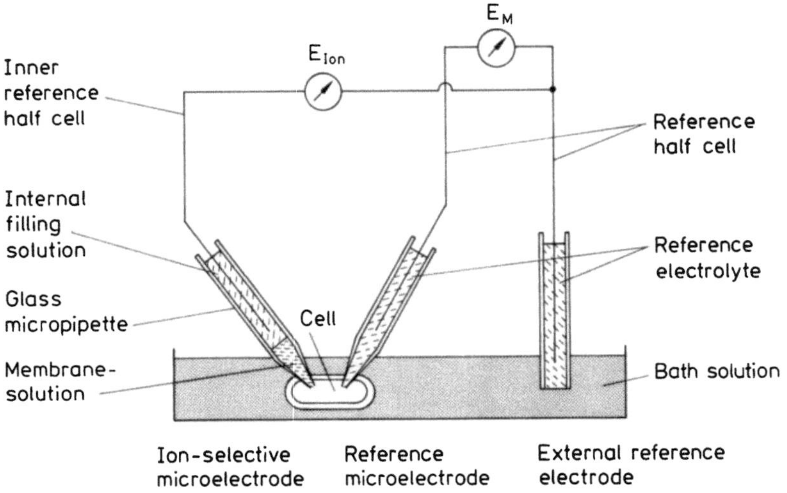

Fig. 5. Schematic diagram of microelectrode arrangement for intracellular measurements of ion activity. (From 120, with permission)

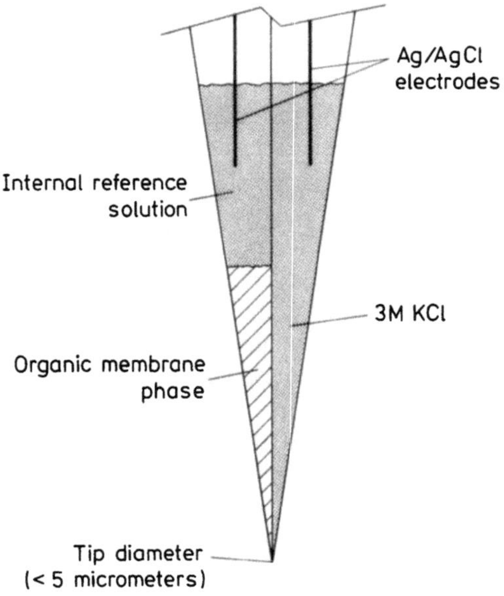

Fig. 6. Double-barrel micropipet ISE for measuring transient intracellular ion activities. (From 11, with permission)

H^+ [129] and acetylcholine [130]. Perhaps the most widely used of these are the liquid membrane electrodes [131,132], which exhibit response characteristics comparable to their conventional-sized analogues.

Microelectrodes can be made sufficiently small to measure intracellular ion activities, as shown schematically in Fig. 5 [120]. The micro ion-selective electrode is inserted into a single cell by means of a micromanipulator. The intracellular activity of the

Table 3. Some representative intracellular applications of micro-ISEs[a]

Cell type	Electrode type	Analyte ion activity	Ref.
Frog heart ventricle	Cl^- liquid membrane micropipet (Corning exchanger #477915)	Cl^-; 17.6 ± 0.57 m*M*	136)
	K^+ liquid membrane micropipet (valinomycin)	K^+; 86.2 ± 0.65 m*M*	137)
Rabbit heart muscle sarcoplasm	Ca^{2+} liquid membrane micropipet (neutral ligand — ETH 1001)	Ca^{2+} (resting muscle) 38 ± 17 n*M*	138)
Neutrons of the molluse, *Aplusia californica*	Ca^{2+} liquid membrane micropipet (ion-exchanger type with PVC added)	Ca^{2+}; 540 n*M*	139)
Vacnoles of plant cells (*Acer pseudoplatanus*)	H^+ liquid membrane micropipet (tridodecylamine as neutral ligand)	pH; 6–6.5	140)
Renal tubules of rats	K^+ liquid membrane micropipet — double-barrel design (valinomycin)	K^+ (proximal tubule) 54.4 ± 2.5 m*M*	141)
Renal tubules of bull frogs	Cl^- liquid membrane micropipet (ion-exchanger type)	Cl^-; 9.2 m*M*	142)
Giant snail neurons	Ca^{2+} liquid membrane micropipet (alkyl phosphate ion-exchanger type)	Ca^{2+}; 450 n*M*	143)
Squid axons (*Lollgo pealii*)	H^+ sealed-end all-glass microelectrode	pH; 7.3	144)
Muscle fiber of frog satorius	K^+ liquid membrane micropipet — double-barrel design (valinomycin as carrier)	K^+; 105 m*M*	145)
Giant neuron of *Aplysia californica*	N_2^- closed-end all-glass microelectrode	Na^-; 40.3 m*M*	146)
Sheep heart Purkinje fibers	N_2^- closed end, recessed up — all-glass microelectrode	Na^-; 7.3 m*M*	147)
	Cl^- liquid membrane micropipet (ion-exchanger based)	Cl^-; 13.8 m*M* (quiescent state)	148)
Chironomus parvae salivary cland cells	Na^- liquid membrane micropipet [ion-exchanger based on tetra(-*p*-chlorophenyl)borate]	Na^-; 16 m*M*	149)

From [11], with permission

analyte ion is then monitored by measuring the potential difference between the ISE and a reference electrode. Another reference microelectrode can be placed in the bathing solution outside the cell, and the potential between the two reference electrodes measured in order to simultaneously monitor the membrane potential of the cell. A microelectrode of the "double-barrel" micropipet design is shown in Fig. 6 [11, 133, 134]. One barrel-tip contains the organic membrane phase and an internal reference electrode; the other constitutes a second reference electrode. A four-barrel configuration with a 1-µm tip in which three barrels are liquid membrane electrodes for Na^+, Ca^{2+} and K^+ and the fourth is a reference electrode has been reported [135]. Some representative applications of ion-selective electrodes for intracellular measurements are shown in Table 3.

Catheter-type ion selective electrodes have been developed for continuous in vivo monitoring of ion activities and blood gases [11]. A miniature ion selective and reference electrode are incorporated in a catheter tube (o.d. ca. 2 mm or less) which is then implanted. Ion-selective electrodes of this type for in vivo measurements are still in the experimental stage. Problems that are being addressed include noise encountered during in vivo measurements, electrode drift, blood clotting, reduction of junction potential errors, and the development of adequate standards for calibration purposes.

An alternative approach to implantation of the ISE is the continuous on-line monitoring of small volumes of blood that are continuously removed from the patient. Dilution of the blood within the instrument prior to making the measurement eliminates junction potential, ionic strength and clotting problems. A "Biostator" instrument for continuously monitoring Ca^{2+} and K^+ with ion selective electrodes and glucose by an amperometric device (vide infra) during surgical procedures has been developed [150].

2.6 Potentiometric Immunoassay

The remarkable selectivity that is inherent in the reaction of an antibody with the antigen or hapten against which it was raised is the basis for the extensive use of immunoassay for the rapid analysis of samples in clinical chemistry. Immunochemical reactions offer a means by which the applicability of potentiometric techniques can be broadened. A number of strategies for incorporating immunoassay into the methodology of potentiometry have been explored [151].

An electrode in which an antibody or an antigen/hapten is incorporated in the sensing element is termed an "immunoelectrode". The potential response of the immunoelectrode is based on an immunochemical reaction between the sensing element of the electrode and antibody or antigen/hapten in the sample solution. One example of such an electrode is the polymer membrane electrode shown in Fig. 7. The selective response of this electrode to specific immunoglobulins is based on the interaction between antibody in solution and an antigen-ionophore complex in the membrane [152–154]. The electrode is exposed to a constant activity of a marker ion (K^+ in this example) chosen for its compatibility with the ionophore portion of the conjugate. This gives a stable, reproducible background potential. The addition to the sample solution of antibody for the antigen portion of the conjugate causes a potential change in proportion to the concentration of antibody. A detection limit in the µg/mL range has been

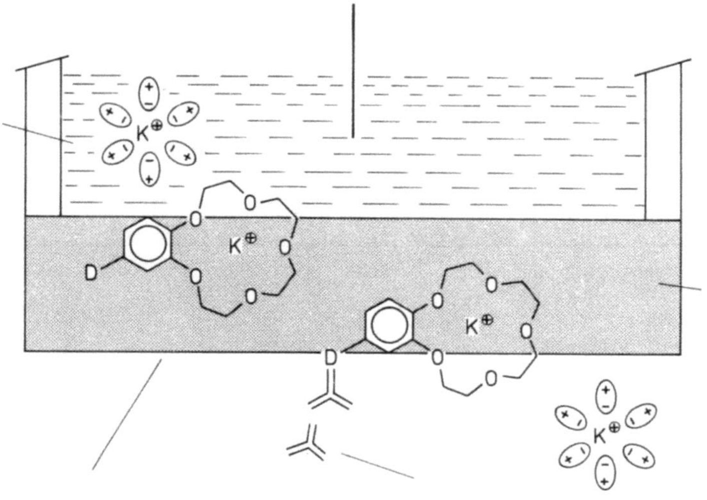

Fig. 7. Schematic diagram of digoxin antibody sensing electrode: (a) PVC membrane containing digoxin-carrier conjugate; (b) inner filling solution, 0.01 M KCl; (c) plasticizer, dibutyl sebacate; (d) digoxin antibodies. (From 152, with permission)

demonstrated for digoxin antibody [152]. An electrode of this type can also be based on a proton carrier [155]. Immobilizing an antigen in a membrane without benefit of an ion carrier also gives a potential response to specific antibody as demonstrated with a complex of cardiolipin antigen and Wassermann antibody [156].

Coating the antigen or antibody directly on appropriately modified metal substrates has yielded electrodes that respond potentiometrically to antigen-antibody reactions [157–160]. An immunoelectrode has been prepared by coating the gate of a CHEMFET with antibody [161]. It has been shown that the immunological coupling response of some of these electrodes might be a minor component of the overall response, which would make these sensors difficult to use as immunoelectrodes [162]. In general, these electrodes as yet have insufficient sensitivity for most practical immunoassays.

Conventional ion-selective electrodes have been used as detectors for immunoassays. Antibody binding measurements can be made with hapten-selective electrodes such as the trimethylphenylammonium ion electrode [163]. Enzyme immunoassays in which the enzyme label catalyzes the production of a product that is detected by an ion-selective or gas-sensing electrode take advantage of the amplification effect of enzyme catalysis in order to reach lower detection limits. Systems for hepatitis B surface antigen [164, 165] and estradiol [166] use horseradish peroxidase as the enzyme label and iodide electrode as the detector. The horseradish peroxidase catalyzes the oxidation of p-fluoroanadine with the fluoride detected by the ISE. Biotin [167] and cyclic AMP [168] have been determined using lysozyme and urease as labels. The immunoreaction between human antibody (IgG) and peroxidase-labeled anti-human IgG antibody can be detected with a fluoride electrode [169]. Adenosine deaminase, asparaginase and urease have been examined as possible enzyme labels for immunoassays using potentiometric detection with the ammonia gas-sensing membrane electrode [170] and adenosine deaminase with an NH_4^+ electrode [171]. CO_2 gas-sensing electrodes have been used for the determination of human IgG [172] and digoxin [173].

Immunoassays have been based on the potentiometric measurement of marker ions such as tetrapentylammonium ion (TPA^+) that are loaded in phospholipid liposomes [174-176]. Complement mediated immunolysis of these loaded vesicles is caused by the presence of appropriate antibodies. The amount of marker released is related to antibody concentration. Similarly, complement and antibody levels have been measured using marker (trimethylphenylammonium, $TMPA^+$)-loaded sheep red blood cell ghosts in which released $TMPA^+$ is measured with a $TMPA^+$ ion selective electrode [177].

3 Dynamic Techniques

Dynamic techniques are those in which electrolytic processes occur at the electrodes and therefore a finite current is passed through the electrochemical cell. This discussion will be limited to controlled-potential techniques, namely voltammetry and amperometry. While other dynamic electrochemical techniques have been developed, these two are by far the most commonly used for bioelectroanalytical studies.

All of the controlled-potential electrochemical techniques are performed by applying some potential waveform to an electrode immersed in the sample solution. The different techniques are achieved either by varying the excitation potential waveform or by changing the monitored response domain. This response can be current, charge, or in some cases spectral in nature. An advantage of electrochemical techniques is that the instrumental requirements are rather simple and essentially the same for all techniques. A basic system consists of a potentiostat for potential control of an electrode and a means of detecting the response (usually the current through the electrode). Indeed, single instruments capable of performing many electrochemical techniques are available from several sources.

A fundamental concept essential to understanding the processes occurring during electrochemical experiments, and therefore necessary for the proper application of these techniques, is the surface nature of electrochemistry in electrolyte solutions. All redox reactions occur in the interphase region between the bulk solution and the electrode surface. When a potential difference is applied between two electrodes in an electrolyte solution the potential gradient exists only in the narrow interphase regions near the two electrodes, with the bulk of the solution remaining at electroneutrality. Therefore, molecules in the bulk solution cannot feel the presence of the electrodes or the potential gradients. Only molecules within the potential gradient (typically less than 10^{-6} m from the electrode) are probed during an electrochemical experiment.

It is also important to understand how the potential gradient between an electrode and the bulk solution is established and controlled. Because the potential difference between the electrode and the bulk solution is not measurable, a second electrode must be employed. Although in general the potential difference between an electrode and solution cannot be determined, the potential difference between two electrodes in that solution can be determined. If the solution electrode potential difference of one of the electrodes is held constant by maintaining a rapid redox couple such as silver-silver chloride or mercury-mercurous chloride (calomel), then the potential difference between the solution and the second electrode can be directly related to the potential difference between the two electrodes. Although the potential difference between the electrode and solution is not known in an absolute sense, it is established in a relative sense.

In the two electrode system described above the electrode of maintained potential difference is known as the counter electrode while the electrode at which the potential difference is established relative to the counter electrode is termed the working electrode. It is at the working electrode where the electrolytic processes of interest to electroanalytical chemistry occur. The counter electrode serves the second purpose of completing the electrical circuit and allowing charge to pass through the cell. The two roles of the counter electrode are not independent because a current through the electrode can cause a change in its potential difference with the solution. This problem is circumvented by splitting the roles of the counter electrode between two electrodes to give an overall three electrode system. In the three-electrode system a reference electrode is used to maintain a constant solution potential difference, an auxiliary electrode is used to complete the electrical circuit, and a working electrode is used to apply a potential difference to drive electrolytic reactions (Fig. 8). In this arrangement there is negligible current through the reference electrode so its solution potential difference is constant regardless of the current through the electrochemical cell (working to auxiliary electrode).

In a high dielectric constant medium (i.e. water), when a potential gradient is established between an electrode and the solution a charge excess will develop on the electrode surface which must be balanced by solution species. Solvent molecules and other species (i.e. electrolyte) orient themselves at the electrode surface to counter the elec-

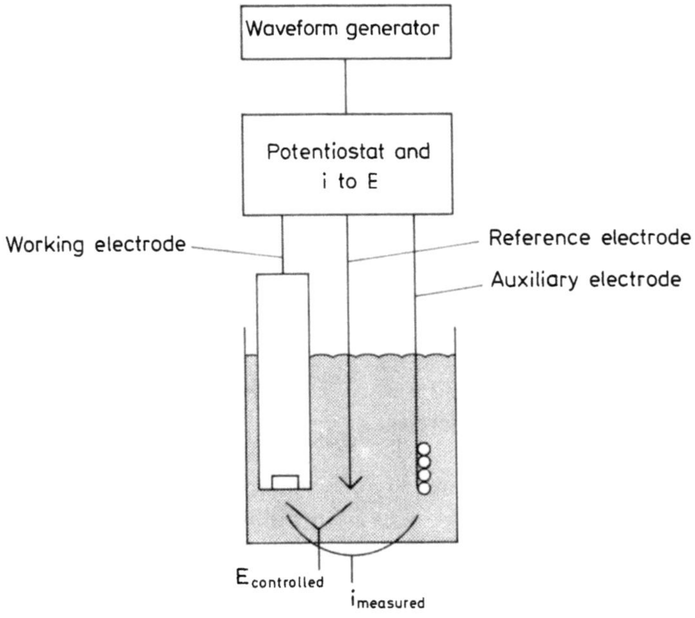

Fig. 8. Three electrode voltammetry system

trode charge. This region is known as the double-layer because of the arrangement of molecules thought to exist. This electrode-solution, double-layer region behaves in a manner similar to a capacitor. When a potential difference is applied at the electrode solution interphase, charge will accumulate at the electrode and the amount of charge will be a function of the potential difference applied. "Double-layer charging currents" are thus produced when the potential difference is changed. The significance of these charging currents will be discussed later with regard to voltammetric techniques.

All Faradaic (electrolytic) electrochemistry (as the name implies) ultimately depends on Faraday's Law:

$$Q = nFN \tag{8}$$

where Q is the charge (coulombs), n is the number of electron transferred (equivalents/mole), F is the Faraday constant (96,485 coulombs/equivalent), and N is the number of moles of reactant. Faraday's law is the most fundamental concept in electroanalytical chemistry because it directly relates the amount of chemical reaction to an electrical response. For analytical purposes it is often more useful to consider the integrated from of Faraday's Law:

$$i_t = (dQ/dt)_t = nF(dN/dt)_t \tag{9}$$

where i_t is the current (amperes) at time t and dN/dt is the rate of conversion of reactant at the electrode surface. In this form, Faraday's Law becomes a rate equation relating the rate of a chemical reaction to an electrical response. Electrochemistry is one of the few techniques which makes instantaneous, single point rate measurements. The equations describing the responses for the various electrochemical techniques are all based upon this equation.

Consideration of Faraday's law shows the advantage of electrochemistry for analytical purposes. With currents as small as 10^{-9} A easily measurable, conversion rates of as little as 10^{-14} eq/s are readily detectable. This means detection limits in the range of 10^{-9} M can be achieved by electroanalytical techniques!

While electroanalytical techniques are inherently quite sensitive, the resolution of a mixture of electroactive compounds is very difficult. Practical considerations limit the usable "potential window" to no more than 3 V and typically around 1.5 V. This is because at more extreme potentials the medium or the electrode itself begin to oxidize or reduce. In addition, the electrochemical response of compounds as a function of applied potential is fairly broad so that at least a 200–400 mV difference in half-wave potentials is required for adequate resolution. This typically limits electrochemical resolution of mixtures to no more than three or four electroactive compounds.

Because of this lack of resolving power, much electroanalytical research is aimed at providing increased selectivity. This can be accomplished in two ways. First, electrochemistry can be combined with another technique which provides the selectivity. Examples of this approach are liquid chromatography with electrochemical detection (LCEC) and electrochemical enzyme immunoassay (EEIA). The other approach is to modify the electrochemical reaction at the electrode to enhance selectivity. This approach is exemplified by modified electrode methods where reaction at the electrode surface is limited beyond mere electrochemical considerations to include physical and chemical properties. The following discussion will illustrate in detail how these approaches can provide analytical techniques with both high selectivity and low detection limits.

3.1 Liquid Chromatography/Electrochemistry

Liquid chromatography/electrochemistry (LCEC) has become recognized as a powerful tool for the trace determination of easily oxidizable and reducible compounds. This is because detection of as little as 0.1 pmol of material is readily accomplished with relatively simple and inexpensive equipment. Initial interest in LCEC was generated by the determination of several aromatic matabolites of tyrosine in the central nervous system. However, the application of LCEC into other areas of biochemistry has begun at a growing pace. A bibliography of LCEC applications is available [178].

3.1.1 Hydrodynamic Voltammetry

LCEC is a special case of hydrodynamic chronoamperometry (measuring current as a function of time at a fixed electrode potential in a flowing or stirred solution). In order to fully understand the operation of electrochemical detectors, it is necessary to also appreciate hydrodynamic voltammetry. Hydrodynamic voltammetry, from which amperometry is derived, is a steady-state technique in which the electrode potential is scanned while the solution is stirred and the current is plotted as a function of the potential. Idealized hydrodynamic voltammograms (HDVs) for the case of electrolyte solution (mobile phase) alone and with an oxidizable species added are shown in Fig. 9. The HDV of a compound begins at a potential where the compound is not electroactive and therefore no faradaic current occurs, goes through a region

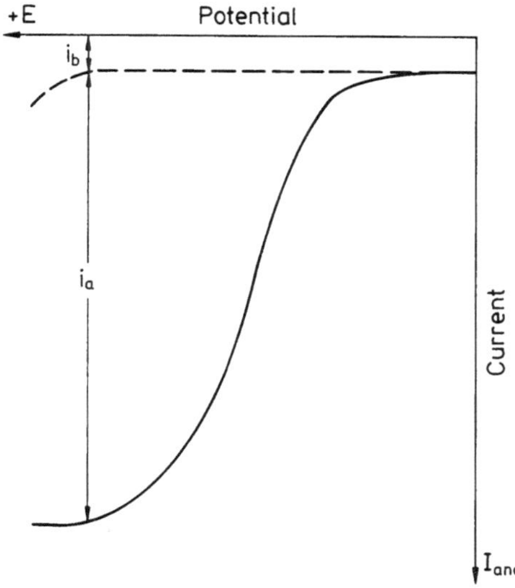

Fig. 9. Generalized hydrodynamic voltammogram. i_b = background current, i_a = Faradaic current from analyte oxidation

of increasing current, and finally reaches a limiting current plateau, where essentially all of the compound reaching the electrode surface is oxidized. The HDV is characterized by the half-wave potential, $E_{1/2}$, which is defined as the potential at which the current is one-half its limiting value. The slope of the rising portion of the HDV is determined by the Nernst equation and the kinetics of electron transfer at the electrode surface.

The hydrodynamic voltammogram is used to select the operating potential for an LCEC experiment. Two considerations are of major importance. First, quantitative determinations are based on the degree to which the redox current from the analyte can be distinguished from the background current. As the background current typically increases with increasing potential, this means operating at the smallest reasonable applied potential. Second, operating on the limiting current plateau means that the redox current from the analyte is independent of applied potential. Because changes to the electrode surface during the course of an LCEC experiment can cause small changes in the interphase potential difference, operating on the limiting current plateau assures a constant response even if these small variations occur. Overall, these considerations dictate that a potential just on the limiting current plateau is normally chosen for LCEC experiments.

3.1.2 Mobile Phase Considerations

While all detectors place some limitations on the mobile phase composition, in electrochemical detection it is essential to recognize that a complex surface reaction is involved which depends on both the physical and chemical properties of the medium. To optimize an LCEC determination, it is necessary to consider both chromatographic and electrochemical requirements simultaneously. The primary requirement for electrochemical detection is that the mobile phase have a relatively high conductivity. To this end, buffered mobile phases of moderate ionic strength (0.01–0.1 M) are typically used. Obviously, totally nonpolar organic mobile phases cannot be used because of their inability to support a significant ionic strength. This rules out classical normal phase separations. The vast majority of LCEC applications have employed reverse phase, ion exchange, or ion-pair separations. While aqueous in nature, significant amounts (up to 90% v/v) of organic modifier can be added to the mobile phase. Totally nonaqueous mobile phases can be used if the solvent can support an electrolyte. Organic solvents which have been used with LCEC include methanol, acetonitrile, DMF, and DMSO with salts such as tetrabutylammonium hexafluorophosphate or tetrafluoroborate as the electrolyte.

3.1.3 Electrode Materials

The choice of the electrode material can be critical to the successful use of LCEC for the determination of biochemical molecules. Historically, the material of choice is carbon paste, a mixture of graphite powder with a dielectric binder. Carbon paste exhibits low background currents at positive potentials and good mechanical stability. However, carbon paste electrodes are incompatible with mobile phases containing more than about 20% organic solvent. More recently, glassy (or vitreous) carbon electrodes have become more popular than carbon paste electrodes. This is primarily because glassy carbon is compatible with organic solvents and is easier to use. In addition, glassy carbon typically exhibits lower background currents at negative potentials than does carbon paste. Mercury electrodes have also been employed with LCEC. Because of the poor mechanical stability of mercury alone, it is often amalgamated with gold to form a Au/Hg amalgam electrode. Mercury has the best characteristics when operating at negative potentials, but it is generally not applicable to oxidative detection because of the low oxidation potential of mercury itself. Platinum electrodes have been employed to a limited extent but have found little acceptance due to the problem of surface oxide layer formation. It is impossible to say that one electrode material will be superior to another in all situations. It is best to evaluate each redox system of interest with several types of electrodes to make the optimal choice.

3.1.4 Cell Design

The design of the electrochemical flow cell can dramatically affect the performance of the detector. For this reason several different cell designs have been devised for chromatographic detection. The most popular design is the thin-layer cell in which the working electrode is part of one wall of a thin-layer channel through which the chromatographic eluent flows. Thin-layer cells can be constructed such that flow is parallel to the electrode [179] or directed perpendicular to the surface followed by radial dispersion (known as the wall-jet cell) [180] as shown in Fig. 10. To date, conventional thin-layer cells have shown performance superior to wall-jet cells primarily due to better noise characteristics. The thin-layer cell design provides a high ratio of electrode surface area to solution volume providing high sensitivity and excellent flow characteristics of well developed laminar flow for low flow related noise. In addition, the thin-layer cell design is compatible with essentially all electrode materials and offers convenient access to the electrode for resurfacing. Most commercially available electrochemical detectors use thin-layer cells.

Flow through electrochemical detectors based on a cylindrical geometry, as opposed to a planar geometry, have also been developed [181–184]. Three cell designs using cylindrical geometry have been used with liquid chromatography (Fig. 10). Electrochemists like open tubular electrodes because the mathematics of mass-transfer in a cylindrical geometry are relatively easily solved. From a practical perspective, open tubular electrodes are less attractive because many common electrode materials can not be easily used (e.g. mercury and carbon paste) and resurfacing of electrodes is

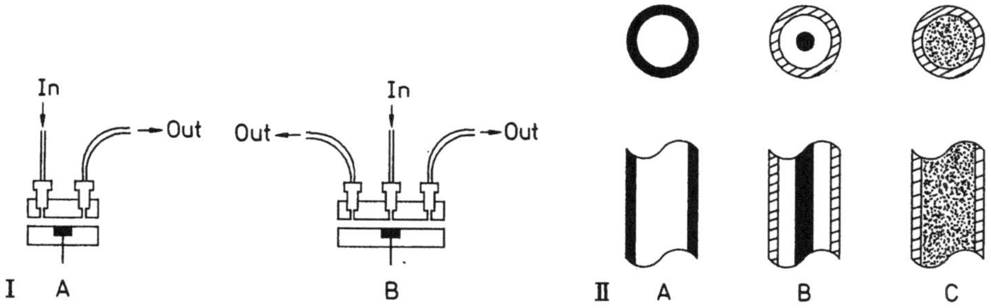

Fig. 10. Flow-through electrochemical cell designs. I, Planar geometries, thin-layer (A) and wall-jet (B) flow cell designs. II, Cylindrical geometries, open tubular (A), wire in a capillary (B), and packed-bed (C) flow cell designs

difficult. The surface to volume ratio is also not very large. The wire in a capillary cell design has become more attractive with the advent of open tubular liquid chromatography columns. Jorgenson and coworkers [185] have described the fabrication of such an electrochemical cell directly at the end of an open tubular chromatographic column.

Packed-bed electrodes (Fig. 10) are useful for constructing electrochemical detectors of high conversion efficiency. Because typical packed-bed cells have very large electrode surface areas, high background currents are a problem. Another disadvantage of packed-bed cells is the difficulty of resurfacing the electrode. It is not possible to physically resurface the electrode by polishing and less vigorous chemical methods must suffice. Tubular packed-bed electrode cells are the best choice for electrosynthetic applications where 100% conversion is important. A packed-bed cell is available with a commercial electrochemical detector.

Most LCEC applications use a single working electrode, although it is possible to monitor the current at several electrodes simultaneously. In practice this becomes more difficult as more working electrodes are employed so that practical applications have been limited to two working electrodes. Several advantages can be achieved using two working electrodes relative to single electrode detection [186]. So-called "dual-electrode" detectors are commercially available in both the thin-layer and packed-bed designs.

Both cell designs permit positioning of the second electrode downstream of the first working electrode (Fig. 11), which is known as the series configuration. This electrochemical transducer is used in the same manner as the classic ring-disk electrode. Products generated at the upstream electrode are detected (or collected) at the downstream electrode [187,188]. Selectivity is enhanced when the products of the upstream electrolysis can be detected at a more favorable potential than is necessary for the upstream reaction. In addition, chemically irreversible redox couple are not detected downstream in the series configuration further enhancing selectivity and improving detection limits. The series configuration has also been used to study the electrochemical reactions occuring at the upstream electrode [189].

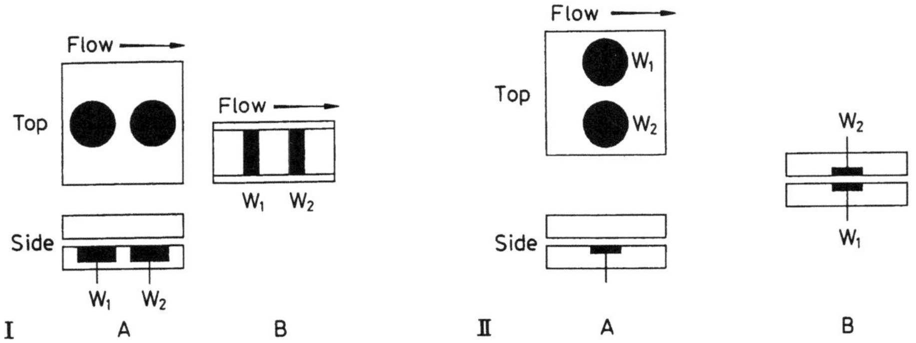

Fig. 11. Dual-electrode cell configurations. I, series configurations, thin-layer (A) and packed-bed (B) cell designs. II, parallel configurations, parallel adjacent (A) and parallel opposed (B) cell designs

The thin-layer cell design also permits two electrodes to be placed adjacent to each other and normal to the flow path (Fig. 11), which is known as the parallel adjacent configuration. With this configuration the current at each electrode (typically at different applied potentials) is monitored individually and is analagous to dual-wavelength UV absorption detectors. The parallel configuration can be used in three ways. First, both oxidizable and reducible compounds can be detected simultaneously [190]. Second, easily oxidized (or reduced) analytes can be selectively detected in the presence of hard to oxidize (or reduce) analytes by using one electrode at a relatively low potential to detect the easily oxidized analytes and using the other electrode to detect the hard to oxidize analytes at a less selective potential [191]. Finally, by operating both electrodes at different potentials on the HDV of a compound, voltammetric data can be obtained from a single chromatographic analysis providing qualitative information [191–193].

A third configuration of the dual-electrode thin-layer cell is also shown in Fig. 11. This geometry is known as the parallel-opposed configuration and provides enhanced response for reversible compounds by redox cycling. As the analyte is cycled from its oxidized to reduced form, many more electrons are transferred than would be with a single working electrode and the current response is thus greatly amplified. Unfortunately, with conventional LC systems the volume flow rate through the electrochemical cell is too high to permit a significant number of cycling steps to occur. Goto et al. [194] have been able to achieve up to a 20-fold increase in response using 1 cm long electrodes spaced 30 μm apart with a flow rate of 20 μL/min with a microbore chromatography column. With the growing interest in microbore LC and the associated lower volume flow rates, this concept should find more practical application.

While most LCEC experiments directly use the current response at the electrode for quantitation, it is possible to combine the response from two electrodes into a difference response. Lunte et al. [195] have described advantages for LCEC which can be achieved using the series and parallel configuration of the dual-electrode detector in the difference mode. With the parallel configuration the electrodes are operated at potentials bracketing the region where the current response is most dependent on potential (i.e. $E_{1/2}$). The difference response from the analyte is not significantly different from the response at a single electrode operated at the higher potential. However, the difference response from compounds both harder and easier to oxidize are greatly attentuated. In essence, difference mode detection with the parallel configuration establishes a "potential window" through which only compounds whose current response changes can be seen. The major sources of baseline interference in LCEC, such as flow rate and temperature fluctuations, static discharge, and changes in mobile phase composition are common-mode (i.e. occur at both electrodes simultaneously). Difference mode detection can be an effective method to improve signal quality because these types of interference signals are greatly attenuated in the difference signal.

With the series configuration, both electrodes are operated at the same potential and on the limiting current plateau of the analyte. Because the diffusion layer is depleted of the analyte at the upstream electrode, the downstream response is greatly attenuated and the difference response not greatly different than the upstream response alone. However, signal from baseline interferences is greatly attenuated for a large improvement in the signal-to-noise ratio. Because this mode is based on mass transport instead of electrochemistry, no change in selectivity relative to single electrode operation is achieved.

3.1.5 Conversion Efficiency

Depending upon the cell design and flow rate, electrochemical detectors can electrolyze from ca. 5 % up to nearly 100 % of the material entering the cell (See Footnote). In practice, high efficiency is obtained by using a large surface area electrode. As the electrode are is incresed, each added increment contributes proportionately less to the total signal, but approximately equal amounts to the background. As background noise is roughly proportional to background current, the signal-to-noise ratio (see below for the significance of signal-to-noise to electrochemical detection) tends to decrease as the electrode area becomes very large. Therefore, although the absolute response of an analyte is larger for high efficiency detectors, detection limits are typically no better and often worse than for detectors with a conversion efficiency of 10–20 %. A detector of 100 % conversion efficiency can be useful for quantitation of an analyte for which no standard is available. If the number of electrons involved in the electrochemical reaction is known and the chromatographic peak is baseline resolved so that an accurate peak area can be determined then quantities can be calculated directly from Faraday's Law. This is not a very common occurrence in practice.

3.1.6 Performance Evaluation

With the multitude of transducer possibilities in terms of electrode material, electrode number, and cell design, it becomes important to be able to evaluate the performance of an LCEC system in some consistent and meaningful manner. Two frequently confused and misused terms for evaluation of LCEC systems are "sensitivity" and "detection limit". Sensitivity refers to the ratio of output signal to input analyte amount generally expressed for LCEC as peak current per injected equivalents (nA/neq or nA/nmol). It can also be useful to define the sensitivity in terms of peak area per injected equivalents (coulombs/neq) so that the detector conversion efficiency is obvious. Sensitivity thus refers to the slope of the calibration curve.

Sensitivity by itself is not sufficient to completely evaluate an LCEC system for analytical purposes. The minimum detectable quantity (detection limit) is of more practical importance. The detection limit takes into consideration the amount of baseline noise as well as the response to the analyte. The detection limit is then defined as the quantity of analyte which gives a signal-to-noise ratio of three (a S/N of 3 is the generally accepted criterion although other values have been used). For a complete description of an LCEC application, both the sensitivity and detection limit, along with the S/N criteria used, should be provided.

While the terms "amperometric detection" and "coulometric detection" have come into use to describe detectors of less than 100 % efficiency and 100 % efficiency respectively, these terms are actually misnomers. An amperometric detector is any electrochemical detector where current is plotted as a function of time, regardless of the conversion efficiency. A coulometric detector is any electrochemical detector where charge is plotted as a function of time, again regardless of the conversion efficiency. Preferred terminology should be "high efficiency" and "low efficiency" detectors to describe the two situations.

3.1.7 Applications

LCEC has become a widely used analytical technique for biomedical analysis. Several reports appear each month describing new LCEC analyses or reporting results obtained with LCEC. While space does not permit a complete review of all biochemical applications, this section will consider applications of LCEC to general classes of compounds to provide an overview of the uses of this technique. For a thorough survey of the literature, a bibliography of LCEC applications is available [178].

3.1.7.1 Oxidative Applications

Phenols:
Most phenolic compounds are readily oxidized at carbon electrodes. The oxidation potentials vary widely depending upon the number of ring hydroxyl groups and their positions on the ring. Many compounds of biomedical and industrial interest are phenolic and LCEC based trace determinations are quite popular.

The first and still most common LCEC application is determination of the catecholamines in biological samples. The number of papers describing methods for catecholamine determination for certain circumstances are far too numerous to cite, however, several good reviews are available [196-198]. The second major use of LCEC is in the determination of the hydroxyindole metabolites of tryptophan. Again several LCEC methods for determination of tryptophan metabolites have been reported [199-201]. A recent paper by Seegal et al. [202] describes a general method to determine biogenic amines and metabolites in brain tissue, cerebral spinal fluid, urine and plasma.

Phenols also constitute a major source of xenobiotic exposure to the body in the form of drugs and environmental pollutants. Oxidative metabolism of these compounds can lead to physiological damage, therefore the metabolism of these compounds is of great interest. LCEC has been a powerful tool for investigating the metabolism of aromatic compounds by the cytochrome P-450 system [203-206]. LCEC also provides a powerful method for the trace determination of phenolic pollutants in environmental samples [207].

Phenolic acids and polyphenols are natural plant constituents which impart flavor and textural components to beverages made from these plants. In order to better understand the role of these easily oxidized compounds in the flavor and stability of beverages, it is necessary to determine them at the low concentrations they occur. LCEC has been shown to be quite effective at these trace determinations [208-210].

Aromatic Amines:
Many pharmaceuticals and environmental pollutants are aromatic amines. Like phenols, this class of compound is generally oxidizable at carbon electrodes. LCEC has been used to study the metabolism of aromatic amines of both environmental and pharmaceutical origin [211-213]. LCEC has also been used for the trace determination of aromatic amines in commercial products [214] and environmental samples [215].

Heterocyclic Compounds:
Many compounds of biomedical interest, both of endogenous and exogenous origin, are heterocyclic in structure. Many of these compounds are electroactive at potentials useful for LCEC analysis. Methods for the determination of both ascorbic acid [216,217] and uric acid [218,219] were developed in the early days of LCEC. The important enzyme

cofactors, the folates [220], the pterins [221-223], biotin [224], and NADH [225] are all electroactive heterocycles which have been determined by LCEC.

Amino Acids and Peptides:
While most amino acids are not electroactive at analytically usable potentials at carbon electrodes, much work is currently directed at general methods of LCEC amino acid detection by electrode surface modification or derivatization of the amino acid. Kok et al. [226] have directly detected amino acids at a copper electrode. Several derivatization methods for amino acids have also been reported [227,228].

The pentapeptides, met- and leu-enkephalin, have been detected in rat striatum tissue by LCEC at a glassy carbon electrode [229]. These peptides can be detected directly because they contain an electroactive tyrosine residue. A series of endorphins, also containing tyrosine, have been detected by LCEC [230].

Thiols:
Thiols are easily oxidized to disulfides in solution, but this reaction occurs only very slowly at most electrode surfaces. However, use can be made of the unique reaction between thiols and mercury to detect these compounds at very favorable potentials. The thiol and mercury form a stable complex which is easily oxidized, in a formal sense it is mercury and not the thiol which is actually oxidized in these reactions. For the LCEC determination of thiols a Au/Hg amalgam electrode is used [231]. Using a series dual-electrode both thiols and disulfides can be determined in a single chromatographic experiment [232,233].

Pharmaceuticals:
Many drugs are electroactive, and as such, have been determined using LCEC. Space does not permit a discussion of the relevance of electrochemical detection to each class of drug. Table 4 lists several compounds of pharmaceutical interest (by therapeutic type and electroactive functionality) which have been determined by LCEC.

3.1.7.2 Reductive Applications

The majority of LCEC applications have used oxidative detection. This is likely because of the perceived difficulties encountered with reductive detection. In particular, dissolved oxygen and trace metal ions must be removed to prevent high background currents. These problems are not difficult to overcome [234] and more applications of reductive detection should appear as this is more generally realized.

Nitro Compounds:
Aromatic nitro and nitroso compounds are easily reduced at carbon and mercury electrodes. Other nitro compounds such as nitrate esters, nitramines, and nitrosamines are also typically easily reduced. The complete reduction of a nitro compound consists of three two-electron steps (nitro-nitroso-hydroxylamine-amine). Since most organic oxidations are only two-electron processes, higher sensitivity is typically found for nitro compounds. Several LCEC based determination of nitro compounds have been reported [235-239].

Table 4. Applications of LCEC to drug analysis

Class	Compound	Structure	Ref.
Analgesic	Acetaminophen	Phenol	[204, 205, 251]
	Codeine	Phenol	[252]
	Naproxen	Methoxy phenol	[252]
	Phenacetin	Aromatic amide Ethoxy phenol	[252]
	Salicyclic acid	Phenol	[252]
	Apomorphine	Phenol	[253]
	Salicylamide	Phenol	[252]
	Ketobemidone	Phenolic	[256]
	Morphine	Phenolic	[254, 255]
Tricyclic	Diazepam	Heterocyclic	[257, 258, 260]
Antidepressant	Nitrazepam	Nitroaromatic, Heterocyclic	[257, 259, 260]
	Chlordiazepoxide	Heterocyclic	[257]
Antibacterial	Amoxicillin	Phenolic	[261]
Adrenergic	Labetol	Phenolic	[262]
Blockers	Mepindolol	Indole	[263]
Antineoplastic	Methotrexate	Heterocyclic	[264]
	Procarbazine Hydrochloride	Hydrazine	[265]
Muscle relaxant	Theophylline	Xanthine	[266, 267]
Antihypertensive	Sulfinalol hydrochloride	Phenolic	[268]
Antibiotic	β-cetoterine	Phenolic	[269]
Antitubercular	Rifampicin	Phenolic	[270]
Antipsychotic	Chlorpromazine	Heterocyclic	[271]
Antiarthritic	Penicillamine	Thiol	[272]

Heterocycles:
Several heterocycles of biomedical interest are reducible. Among these the K vitamins [240, 241] and the pterins [241] have been determined by LCEC. Some heterocyclic pharmaceuticals have also been determined by reductive LCEC [242, 243].

3.1.8 Voltammetric Detection

To obtain more complete voltammetric information than is readily found with amperometric detection, various potential scanning detectors have been developed for liquid chromatography. Simple use of a linear ramp waveform, such as used in linear sweep and cyclic voltammetry, is not very useful for chromatographic detection. Because the signal current of an analyte is relatively small in LC detectors, double-layer charging currents can easily obscure the voltammetric information. In addition, the cell resistance is typically large in cells used for chromatographic detection causing a significant iR drop which distorts the voltammogram.

To overcome these problems, most voltammetric detectors have used pulsed waveforms such as staircase [244], squarewave [245], and differential pulse [246]. The current is sampled at the end of the pulse after the charging current has decayed. In addition, because the charging current is typically the major current source, iR problems are not as severe. Last [247] has described a coulostatic detector based on charge pulses instead of potential pulses which eliminates iR and charging current problems. Even with these pulse techniques, voltammetric detection methods have not been able to achieve the detection limits readily obtained by simple amperometric detectors. White et al. [248] have recently described a voltammetric detector using a fiber microelectrode that exhibited greatly reduced charging currents. Unfortunately, this detector was limited to use with open-tubular chromatographic systems.

Lunte et al. [249,250] have recently developed a unique voltammetric detector using a series dual-electrode cell. In the detector, the upstream electrode's potential is scanned while the downstream detector is operated at a constant potential. The downstream electrode is used to monitor the redox reaction occuring at the upstream electrode without the charging current contributions. In essence the upstream electrode is operated voltammetrically and the downstream electrode operated amperometrically so that the detector has been named a voltammetric-amperometric detector. Detection limits of 10^{-7} M have been reported using this detection scheme.

3.2 Enzyme Linked Electrochemical Techniques

Monitoring enzyme catalyzed reactions by voltammetry and amperometry is an extremely active area of bioelectrochemical interest. Whereas liquid chromatography provides selectivity, the use of enzymes to generate electroactive products provides specificity to electroanalytical techniques. In essence, enzymes are used as a derivatizing agent to convert a nonelectroactive species into an electroactive species. Alternatively, electrochemistry has been used as a sensitive method to follow enzymatic reactions and to determine enzyme activity. Enzyme-linked immunoassays with electrochemical detection have been reported to provide even greater specificity and sensitivity than other enzyme linked electrochemical techniques.

Enzyme linked electrochemical techniques can be carried out in two basic manners. In the first approach the enzyme is immobilized at the electrode. A second approach is to use a hydrodynamic technique, such as flow injection analysis (FIAEC) or liquid chromatography (LCEC), with the enzyme reaction being either off-line or on-line in a reactor prior to the amperometric detector. Hydrodynamic techniques provide a convenient and efficient method for transporting and mixing the substrate and enzyme, subsequent transport of product to the electrode, and rapid sample turnaround. The kinetics of the enzyme system can also be readily studied using hydrodynamic techniques. Immobilizing the enzyme at the electrode provides a simple system which is amenable to in vivo analysis.

A wide variety of enzymes have been used in conjunction with electrochemical techniques. The only requirement is that an electroactive product is formed during the reaction, either from the substrate or as a cofactor (i.e. NADH). In most cases, the electroactive products detected have been oxygen, hydrogen peroxide, NADH, or ferri/ferrocyanide. Some workers have used the dye intermediates used in classical colorimetric methods because these dyes are typically also electroactive. Although an electroactive product must be formed, it does not necessarily have to arise directly from the enzyme reaction of interest. Several cases of coupling enzyme reactions to produce an electroactive product have been described. The ability to use several coupled enzyme reactions extends the possible use of electrochemical techniques to essentially any enzyme system.

3.2.1 Off-Line Techniques

The simplest method of coupling enzymatic reactions to electrochemical detection is to monitor an off-line reaction using FIAEC or LCEC. The enzymatic reaction is carried out in a test tube under controlled conditions with aliquots being taken at timed intervals. These aliquots are then analyzed for the electroactive product and the enzyme activity in the sample calculated from the generated kinetic information.

Because LCEC had its initial impact in neurochemical analysis, it is not surprising that many of the early enzyme-linked electrochemical methods are of neurologically important enzymes. Many of the enzymes involved in catecholamine metabolism have been determined by electrochemical means. Phenylalanine hydroxylase activity has been determined by electrochemically monitoring the conversion of tetrahydrobiopterin to dihydrobiopterin [273]. Another monooxygenase, tyrosine hydroxylase, has been determined by detecting the DOPA produced by the enzymatic reaction [274, 275]. Formation of DOPA has also been monitored electrochemically to determine the activity of L-aromatic amino acid decarboxylase [276]. Other enzymes involved in catecholamine metabolism which have been determined electrochemically include: dopamine-β-hydroxylase [277], phenylethanolamine-N-methyltransferase [278], and catechol-O-methyltransferase [279]. Electrochemical detection of DOPA has also been used to determine the activity of γ-glutamyltranspeptidase [280]. The cytochrome P-450 enzyme system has been studied by observing the conversion of benzene to phenol and subsequently to hydroquinone and catechol [203].

In addition to enzyme activity, the concentration of an nonelectroactive substrate can be determined electrochemically by this technique. By keeping the substrate (analyte) the limiting reagent, the amount of product produced is directly related to the initial concentration of substrate. Either kinetic or equilibrium measurements can be used. Typically an enzyme which produces NADH is used because NADH is readily detected electrochemically. Lactate has been detected using lactate dehydrogenase, and ethanol and methanol detected using alcohol dehydrogenase [281].

Lequea et al. [282] used the activity of tyrosine apodecarboxylase to determine the concentration of the enzyme cofactor pyridoxal 5′-phosphate (vitamin B6). The inactive apoenzyme is converted to the active enzyme by pyridoxal 5′-phosphate. By keeping the cofactor the limiting reagent in the reaction by adding excess apoenzyme and substrate, the enzyme activity is a direct measure of cofactor concentration. The enzymatic reaction was followed by detecting tyramine formation by LCEC. The authors used this method to determine vitamin B6 concentrations in plasma samples.

3.2.2 Enzyme Reactors

By incorporating the entire analytical scheme (enzyme reaction and electrochemical detection) into the flow system a great improvement in precision can be realized. Sample manipulation is minimized because only a single injection into the flow system is required versus sampling of aliquots for the off-line method. Precision is also improved because the timing of the enzyme reaction and detection are much better controlled in the flow system. Finally, less of both enzyme and sample are needed with on-line enzyme reactor methods.

The simplest design for an enzyme reactor is to merely have the substrate and enzyme in two merging buffer streams followed by a reaction delay coil (Fig. 12). The delay

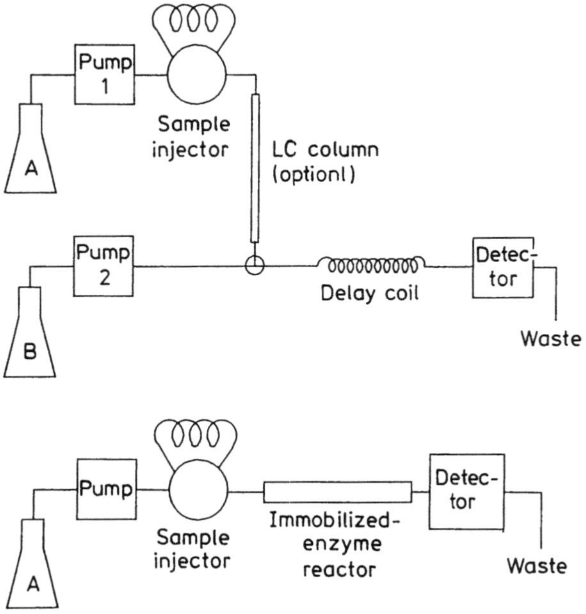

Fig. 12. On-line enzyme reactor system designs, merging stream system (Top) and immobilized-enzyme reactor system (Bottom). A = mobile phase, B = enzyme solution

coil can be cut to the optimum length for complete substrate conversion to product or for any desired percentage of conversion. The electrochemical detector is placed after the delay coil. Throughput is high because, even though a reaction time of several minutes may be required, several samples can be in the flow stream at the same time. Potter et al. [283] have determined choline and acetylcholine using acetylcholinesterase and cholinoxidase to produce hydrogen peroxide. The hydrogen peroxide was detected amperometrically at a platinum electrode. The choline and acetylcholine were separated by liquid chromatography before mixing with the enzymes.

A great savings in enzyme consumption can be achieved by immobilizing the enzyme in the reactor (Fig. 12). In addition to the smaller amount of enzyme required, immobilization often increases the stability of the enzyme. Several designs of immobilized-enzyme reactors (IERs) have been reported, with open-tubular and packed-bed being the most popular. Open-tubular reactors offer low dispersion but have a relatively small surface area for enzyme attachment. Packed-bed reactors provide extremely high surface areas and improved mass transport at the cost of more dispersion.

A wide variety of enzyme systems have been used as IERs with electrochemical detection. The choline and acetylcholine assay has been modified to use a packed-bed IER containing the two enzymes coimmobilized [284]. Phenolic and cyanogenic glycosides have been detected following chromatographic separation by reaction in a packed-bed IER containing β-glucuronidase [285]. Bile acids have been detected using immobilized 3-α-hydroxysteriod dehydrogenase with electrochemical detection of the NADH produced in the reaction [286]. Again, the individual bile acids were chromatographically separated prior to the IER. Ethanol, lactate, and glycerol have all been determined by the amperometric detection of NADH formed as a product of the reaction with immobilized alcohol dehydrogenase, lactate dehydrogenase, and glycerol dehydrogenase, respectively [287].

3.2.3 Enzyme Electrodes

The final method of coupling enzyme reactions to electrochemistry is to immobilize an enzyme directly at the electrode surface. Enzyme electrodes provide the advantages already discussed for immobilization of enzymes. In addition, the transport of enzyme product from the enzyme active site to the electrode surface is greatly enhanced when the enzyme is very near to the electrode. The concept of combining an enzyme reaction with an amperometric probe should offer all of the advantages discussed earlier for ion-selective (potentiometric) electrodes with a much higher sensitivity. In addition, since the response of amperometric electrodes is linear, background can be selected.

Although enzyme electrodes is an extremely important area of bioelectrochemical analysis, this subject is covered in detail in the chapter by Calabrese and O'Connell in this volume and so will not be further discussed here.

3.2.4 Electrochemical Enzyme Immunoassay

Immunoassay is based on the use of an antibody as a selective chemical reagent for an antigen or hapten analyte. Immunoassays are commonly categorized as heterogeneous, in which antibody-bound antigen is separated from free antigen at some point in the procedure, or homogeneous, in which there is no separation step. Most heterogeneous immunoassays are run in a competitive format in which a standard labeled antigen is allowed to compete with standard antigen for a limited number of antibody sites. The ratio of standard to sample antigen which is bound then reflects the concentration of analyte in the sample. Another widely used assay format is the sandwich immunoassay. In this format, the antigen is "sandwiched" between two different antibodies, one of which is labeled for detection. Sandwich assays are more specific than competitive assays because two highly selective antibody reagents are used rather than one, however, the sandwich format is only applicable to antigens large enough to bind two antibodies simultaneously.

To date, most immunoassays have relied on radioisotope labeling for detection. The desire to avoid the use of radioisotopes has led to the development of alternative labels. One of the most successful alternatives has been to use enzyme labels, thus enzyme immunoassays. In enzyme immunoassays, substrate is added following the antigenic binding. Product is then detected and its concentration or rate of formation is a measure of the antigen (analyte) concentration in the sample. Enzyme immunoassays can be extremely sensitive because of the chemical amplification provided by the enzymatic step. Chemical amplification is the passing of a substance through a cycling or multiplication mechanism to generate a large amount of product [288]. In this way, a trace concentration of analyte results in a significantly higher concentration of product which is actually detected but is still related to the original concentration of analyte. Electrochemical enzyme immunoassay is the use of electrochemical techniques to detect the products from an enzyme immunoassay.

3.2.4.1 Heterogeneous Immunoassays

Several heterogeneous electrochemical enzyme immunoassays have been demonstrated. These are based on the enzyme-linked immunosorbent assay (ELISA) technique

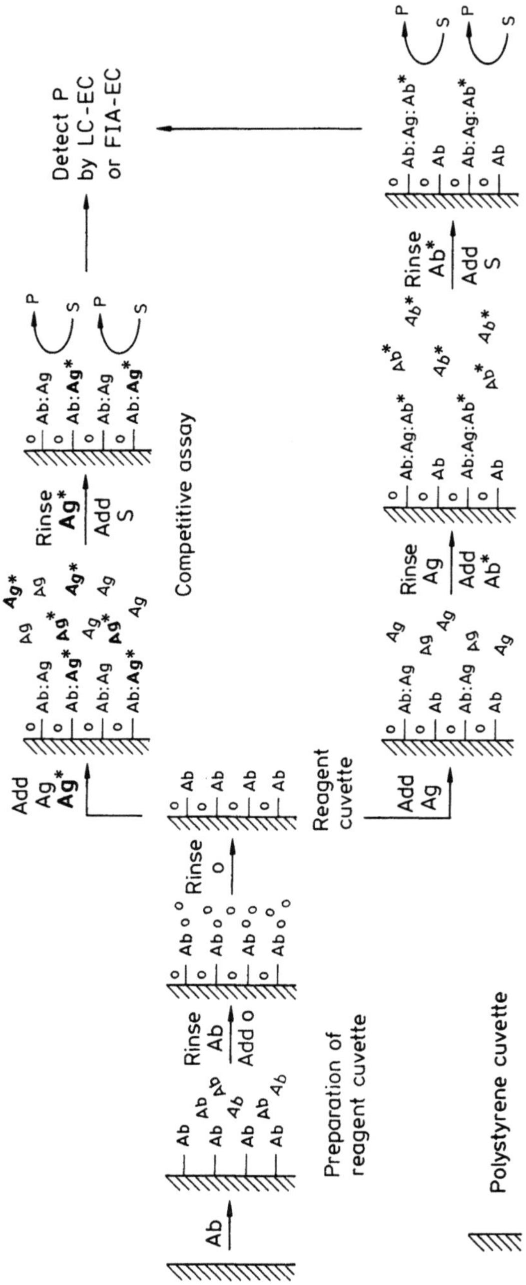

Fig. 13. General protocol for heterogeneous enzyme immunoassay: preparation of reagent cuvettes by coating Ab, competitive assay format, and sandwich assay format. (Reprinted with permission from W. R. Heineman and H. B. Halsall, Anal. Chem. 1985, 57, 1321A. Copyright 1985, American Chemical Society)

in which the antibody is immobilized on the walls of a small-volume plastic cuvette. The ELISA technique can follow either a competitive equilibrium or a sandwich format. Both formats have been used with electrochemical detection. The general protocol for these two formats is shown in Fig. 13.

A competitive electrochemical enzyme immunoassay has been demonstrated for digoxin [289]. Alkaline phosphatase, which catalyzes the hydrolysis of phenyl phosphate to phenol and phosphate, was used as the enzyme label. The phenol which is produced could be detected by either FIAEC or LCEC [289–291]. Although FIAEC has a higher sample throughput, a lower detection limit could be achieved using LCEC because the phenol is separated from the capacitance peak accompanying injection. A detection limit of 50 pg/mL was achieved for digoxin. Competitive electrochemical enzyme immunoassays have also been developed for α_1-glycoprotein [291] and IgG [292].

Catalase has also been used as an enzyme label in competitive heterogeneous enzyme immunoassays. Catalase generates oxygen from hydrogen peroxide with the oxygen determined amperometrically with an oxygen electrode. This approach has been demonstrated for α-fetoprotein [293], theophylline [294], and human serum albumin [295, 296].

A sandwich electrochemical enzyme immunoassay has been described for IgG [292]. Alkaline phosphatase was again used as the enzyme label with the conversion of phenyl phosphate to phenol being determined electrochemically by LCEC. A detection limit of 10 pg/mL was reported.

Heterogeneous electrochemical enzyme immunoassays have several advantageous features. The detection limit is typically in the low pg/mL range, and is a function of the antigen-antibody binding constant, rather than the ability to detect the enzyme product. Since the sample is rinsed from the reagent tubes before adding substrate, problems with interferences by electroactive constituents of the sample or possible electrode fouling by protein adsorption are eliminated. Finally, this technique should lend itself readily to automation for the routine analysis of many samples.

Indeed, automation of a sort has already been accomplished. An electrochemical sandwich immunoassay in which the chromatographic and immunoassay steps are combined in a continuous flow scheme has been described by de Alwis and Wilson [297]. In this technique, antibody is immobilized on a short chromatography column (immunoreactor) analogous to the IERs used for enzyme coupled detection. The antigen sample is injected into the immunoreactor where binding occurs. Enzyme-labelled Ab* is then injected. Finally substrate solution is injected and product detected electrochemically. After analysis an acidic buffer is passed through the immunoreactor to displace the antigen in preparation for the next sample. This technique provides controlled conditioning of the immunoreactor. The immunoreactors are reported to be stable for at least 3 months or 500 assays. The technique was demonstrated for IgG with a detection limit of 1 femtomole. Glucose oxidase was used as the enzyme label with the H_2O_2 produced detected by oxidation at a platinum electrode.

3.2.4.2 Homogeneous Immunoassays

Homogeneous immunoassays rely on a change in the intensity of the label signal that occurs when labeled antigen binds with antibody. When the label is an antibody, a reduction in the rate of enzyme catalysis forms the basis for the assay. This technique

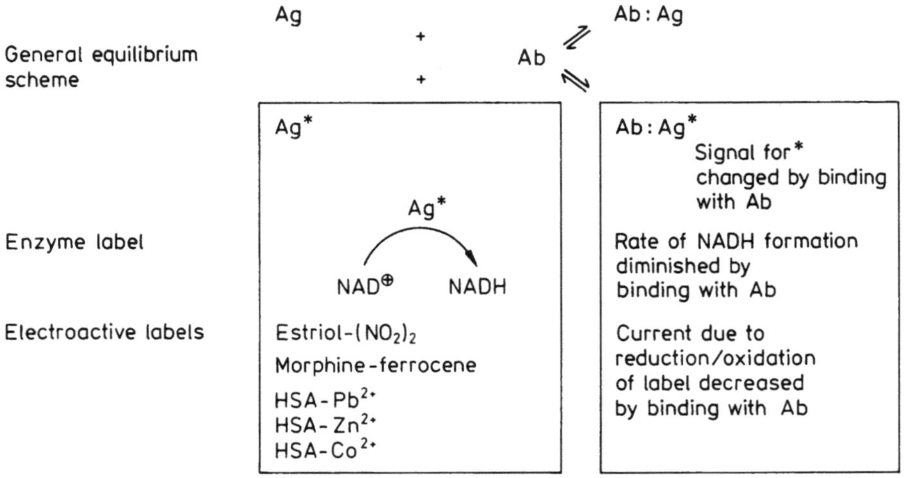

Fig. 14. General reaction scheme for homogeneous immunoassay. (Reprinted with permission from W. R. Heineman and H. B. Halsall, Anal. Chem. 1985, 57, 1321A. Copyright 1985, American Chemical Society)

is commonly known as an EMIT assay (Enzyme Multiplied Immunoassay Technique). This enables free labeled antigen to be distinguished from bound antigen with no separation step necessary. The general scheme for a homogeneous electrochemical enzyme immunoassay is shown in Fig. 14.

Homogeneous electrochemical enzyme immunoassays for both phenytoin [298] and digoxin [299] have been developed. In both cases the label was glucose-6-phosphate dehydrogenase, which catalyzes the reduction of NAD^+ to NADH. The NADH produced was detected by LCEC at a carbon paste electrode.

A homogeneous electrochemical enzyme immunoassay for 2,4-dinitrophenol-aminocaproic acid (DNP-ACA), has been developed based on antibody inhibition of enzyme conversion from the apo- to the holo- form [300]. Apoglucose oxidase was used as the enzyme label. This enzyme is inactive until binding of flavin adenine dinucleotide (FAD) to form the holoenzyme which is active. Hydrogen peroxide is the enzymatic product which is detected electrochemically. Because antibody bound apoenzyme cannot bind FAD, the production of H_2O_2 is a measure of the concentration of free DNP-ACA in the sample.

The main advantage of a homogeneous immunoassay, compared to a heterogeneous immunoassay, is the absence of a separation step. This translates into a simpler procedure and easier automation. However, homogeneous assays are typically less sensitive and more susceptible to sample interferences which are removed in a separation step.

3.3 In vivo Electrochemical Techniques

The first in vivo electrochemical measurements were performed in 1973 [301]. Since then bioelectrochemists have spent much effort in developing in vivo methods of analysis. A major reason for this effort is the rapidity with which in vivo techniques

Table 5. Approximate oxidation potentials of neurochemically important compounds

Compound	Oxidation potential
Dopamine	0.3 V
Epinephrine	0.3 V
5-Hydroxyindole Acetic Acid	0.3 V
5-Hydroxytryptophan	0.3 V
3,4-Dihydroxyphenylacetic Acid	0.3 V
Norepinephrine	0.3 V
Ascorbic Acid	0.3 V
Uric acid	0.4 V
Melatonin	0.5 V
Metanephrine	0.6 V
Normetanephrine	0.6 V
5-Methoxy-4-hydroxyphenylglycol	0.7 V
4,5-Dihydroxyphenylglycol	0.7 V
Homovanillic acid	0.7 V
Tyramine	0.8 V
Octopamine	0.8 V
Tryptamine	0.9 V
Tyrosine	0.9 V
Enkephalins	0.9 V
Tryptophan	1.0 V

can make measurements to follow transient biological processes. Most in vivo electrochemistry has focused on probing neurological processes. This is because the time factor is extremely important in studying neurological systems and because important neurotransmitters, the biogenic amines, are easily oxidized (and therefore easily detected electrochemically). Table 5 indicates some of the compounds which can be detected in vivo by electrochemical techniques and their approximate oxidation potentials. The extensive literature on in vivo electrochemistry reflects the importance of these techniques. An introduction to the literature is contained in a recent text [302].

Electrochemical techniques in vivo use the standard three electrode voltammetric system described earlier with the electrodes implanted in the brain of the animal subject. Measurements are made by acquiring some stable baseline signal and then stimulating release of the biogenic amine neurotransmitters. The change in signal is then a measure of the concentration of neurotransmitter in the extracellular fluid.

3.3.1 Measurement Techniques

3.3.1.1 Chronoamperometry

The most commonly used waveform for in vivo voltammetric measurements is squarewave. This involves the application of a potential pulse to the working electrode for a fixed time at fixed intervals. The current is measured at the end of the potential pulse to minimize capacitive charging current contributions. This waveform is shown in Fig. 15A.

Regularly repeated potential pulses will establish a steady state condition at the electrode surface where diffusion just replenishes the concentration of the compound

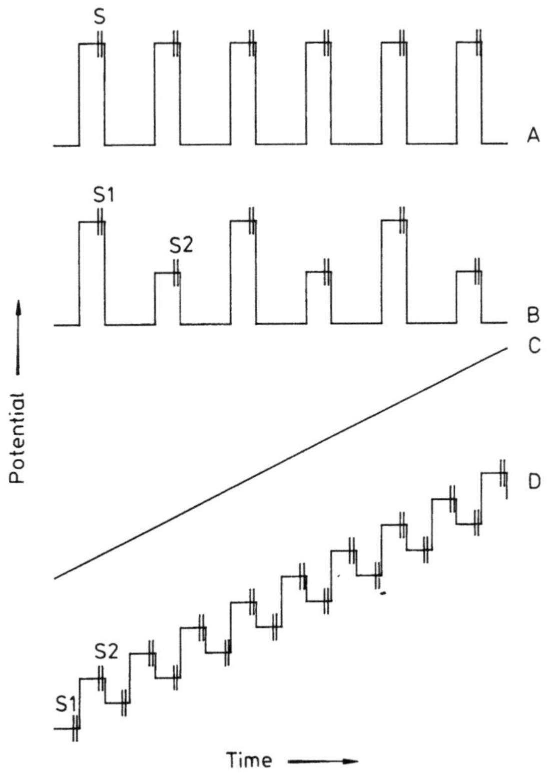

Fig. 15. Waveforms used for *in vivo* electrochemical analysis. A = chronoamperometry, B = double chronoamperometry (response = S1 — S2), C = linear sweep, D = differential pulse (response = S2 - S1), S = sample window

removed by oxidation. It therefore requires several [15–20] initial pulses to reach a steady baseline before useful measurements are made. This is the simplest of the voltammetric techniques but suffers from a lack of selectivity. Any compound which can be oxidized at the pulse potential will be detected providing a summed current response. Because several neurological compounds are easily oxidized, this can make experimental interpretation difficult. Chronoamperometric in vivo analysis has been discussed in detail by Adams and Marsden [303] and be Schenk et al. [304].

To improve the selectivity of chronoamperometric in vivo analysis, a differential measurement technique has been employed [305]. Instead of a single potential pulse, the potential is alternately pulsed to two different potentials giving rise to the name double chronoamperometry. This waveform is shown in Fig. 15B. Because the current contributions of individual electroactive components add linearly to produce the observed current output, the difference in current response at the two potentials is the current due to only those compounds which are oxidized at the higher potential and not oxidized at the lower potential. This system provides two responses, the current due to easily oxidized compounds and the current due to harder to oxidize compounds. This gives greater selectivity than the direct chronoamperometric method.

3.3.1.2 Linear Sweep Voltammetry

Linear sweep voltammetry involves steadily increasing the applied potential and measuring the current response as a function of potential (Fig. 15C). Because voltammetric information about the sample is produced, this technique is much more selective than the chronoamperometric methods. However, the trade off for greater selectivity is a slower analysis time. Because an oxidizing potential is applied for a longer time during the scan than during a potential pulse, it requires a concommitantly longer time to establish steady state conditions. Using a scan rate of 100 mV/sec, Kennett and Joseph were able to collect voltammograms at a rate of only one every 5 minutes [306)]. Using microvoltammetric electrodes Kuhr et al. [307)] were able to perform a scan every second.

3.3.1.3 Differential Pulse Voltammetry

Differential pulse voltammetry has been widely used for in vivo electrochemical analysis [308, 309)]. This technique combines the linear sweep and pulsed potential waveform (Fig. 15D). Current measurements are made just prior to and at the end of each potential pulse. The difference current is plotted as a function of applied potential.

Differential pulse voltammetry provides greater voltammetric resolution than simple linear sweep voltammetry. However, again, a longer analysis time results from the more sophisticated potential waveform. At scan rates faster than 50 mV/sec the improved resolution is lost. Because it takes longer to scan the same potential window than by linear sweep, an even longer relaxation time between scans is required for differential pulse voltammetry.

3.3.2 Electrodes for in vivo Analysis

The most popular material for construction of working electrodes for in vivo analysis has been some form of carbon. The electrodes used in the pioneering studies of Adams and coworkers were made from mixtures of carbon powder in liquid paraffin [310)]. Carbon paste electrodes are easy to construct and exhibit good stability in vivo (up to five days) [311)]. Carbon paste electrodes suffer two major disadvantages. First, they are relatively large (300 μm diameter) for in vivo analysis. Second, carbon electrodes are not selective enough to distinguish many of the easily oxidized neurotransmitters and ascorbic acid (see approximate oxidation potentials in Table 5). The size problem has been partially improved by the use of an epoxy-carbon paste mixture in glass capillaries. These "graphpoxy" electrodes have been made as small as 50 μm in diameter [312)].

The problem of selectivity is the most serious drawback to in vivo electrochemical analysis. Many compounds of neurochemical interest oxidize at very similar potentials. While this problem can be overcome somewhat by use of differential waveforms (see Sect. 3.2), many important compounds cannot be resolved voltammetrically. It is generally not possible to distinguish between dopamine and its metabolite 3,4-dihydroxyphenylacetic acid (DOPAC) or between 5-hydroxytryptamine (5-HT) and 5-hydroxyindolacetic acid (5-HIAA). Of even more serious concern, ascorbic acid oxidizes at the same potential as dopamine and uric acid oxidizes at the same potential as 5-HT, both of these interferences are present in millimolar concentrations

[313, 314]. For these reasons, much work has been done to produce electrodes selective for the catechol neurotransmitters relative to ascorbic acid and uric acid.

Many groups have reported methods of modifying carbon surfaces to make them more selective. Gonon et al. [315] have described an electrochemical pretreatment method consisting of repeatedly applying a triangular potential wave to a carbon fiber electrode immersed in a phosphate buffered saline solution. This pretreatment produces an electrode at which ascorbic acid is more easily oxidized than at the untreated electrode while the oxidation of dopamine and DOPAC is unchanged. Thus dopamine/DOPAC can be determined in the presence of ascorbic acid at an electrochemically pretreated carbon fiber electrode. The disadvantage of this pretreatment technique is that the electrode is stable for only a few hours. A similar electrochemical pretreatment technique has also been reported for "graphpoxy" electrodes [316].

A second surface modification has been reported by Yamamoto et al. [317]. These workers added stearic acid to their carbon paste mixture. This produced an electrode which was relatively insensitive to ascorbic acid and DOPAC relative to dopamine. It is theorized that this electrode works because of electrostatic repulsion of the anionic ascorbate and DOPAC by surface stearate groups. Ionic repulsion has also been employed by covering the surface of the working electrode with an anionic polymer membrane. Gerhardt et al. [318] used Nafion, a hydrophobic sulfonated perfluoropolymer, to make a dopamine selective electrode. This electrode exhibited selectivity coefficients as large as 250:1 for dopamine and norepinephrine over ascorbic acid, uric acid, and DOPAC.

Another approach to improve selectivity is to use an enzyme electrode. The enzyme ascorbate oxidase has been used successfully to remove ascorbate as an interference of in vivo voltammetric electrodes [319, 320]. Ascorbate oxidase converts the ascorbic acid to dehydroascorbate which is not electroactive in the potential region used for in vivo analysis.

Wightman and his coworkers have reported on the use of carbon fiber microelectrodes which provide both a very small electrode and improve the selectivity of the voltammetry. These electrodes have diameters of approximately 20 μm. Because of their small size, these electrodes exhibit lower charging currents and less fouling by electro-generated species than larger electrodes. The authors were able to distinguish ascorbate, dopamine and DOPAC on the basis of the voltammograms obtained [321, 322]. These electrodes were stable for over 8 hours. Another advantage of microelectrodes is that tissue damage due to electrode implantation is minimized.

3.3.3 Conclusion

At present, none of the in vivo electrochemical techniques are capable of measuring basal neurotransmitter levels. Only the increase in extracellular fluid concentrations following some form of stimulation have been measured. The limited selectivity of in vivo electrochemical techniques is also a disadvantage at present. However, the ability to make rapid measurements for continuous sampling to follow the dynamics of neurochemical processes more than offsets the disadvantages of the technique. It cannot be over emphasized that at present only in vivo electrochemical measurements provide a method of monitoring rapid changes in neurotransmitter concentration. Another advantage of in vivo electrochemical techniques is that tissue damage is less severe than with alternative perfusion techniques.

3.4 Anodic Stripping Voltammetry

Anodic stripping voltammetry (ASV) has been used extensively for the determination of heavy metals in samples of biological origin, such as lead in blood. ASV has the lowest detection limit of the commonly used electroanalytical techniques. Analyte concentrations as low as 10^{-10} M have been determined. Figure 16 illustrates ASV for the determination of Pb^{2+} at a mercury electrode. The technique consists of two steps. The potential of the electrode is first held at a negative value for several minutes to concentrate some of the Pb^{2+} from the solution into the mercury electrode as Pb. The electrode process is

$$Pb^{2} + 2e \rightarrow Pb(\emptyset) \tag{10}$$

with the reduction product Pb dissolving in the mercury electrode by amalgamation. During this step the solution is stirred to maximize the rate of accumulation of Pb in the mercury. Thus, lead ion is electrochemically extracted as elemental lead into the mercury electrode, the volume of which is considerably less than the volume of sample solution in the electrochemical cell. The resulting "solution" of metal atoms in the liquid mercury is substantially more concentrated than the solution of metal ions being analyzed, by a factor of up to 10^6.

After discontinuing the stirring, the potential is scanned positively, causing the amalgamated lead to be oxidized back into the solution (i.e., stripped out of the electrode). This oxidation of lead gives a current peak, i_p, the magnitude of which is determined by the concentration of Pb in the mercury electrode which is in turn proportional to the amount of Pb^{2+} in the sample.

During the deposition step, some fraction of the total analyte is deposited into the mercury electrode by electrolysis for a given length of time. An exhaustive electrolysis, in which all of the analyte is deposited into the electrode, is time consuming and generally unnecessary, since adequate concentrations can usually be deposited into

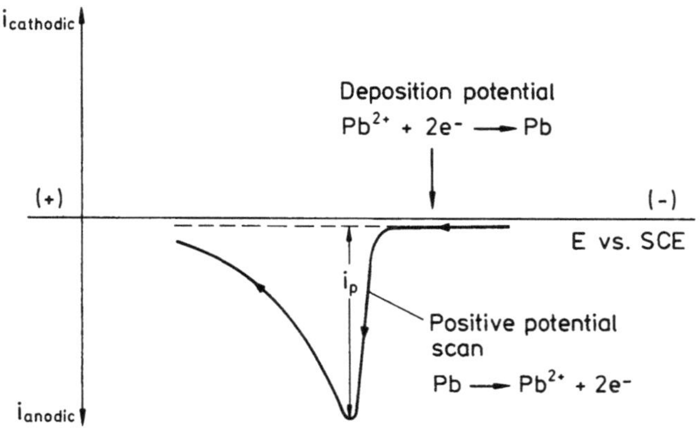

Fig. 16. Anodic stripping voltammogram for Pb^{2+} at the hanging mercury drop electrode. (Reprinted with permission from W. R. Heineman, in "Water Quality Measurement: The Modern Analytical Techniques", (H. B. Mark, Jr. and J. S. Mattson, eds.) Marcel Dekker: New York, 1981)

the electrode to give a satisfactory stripping signal in much shorter times. Since the deposition is not exhaustive, it is important to deposit the same fraction of analyte for each stripping voltammogram in a series of standards and samples. The parameters of electrode surface area, deposition time, and stirring must be carfully duplicated for all standards and samples. Deposition times vary from 60 sec to 30 min, depending on analyte concentration, type of electrode, and stripping technique. The less concentrated solutions require longer deposition times to give adequate stripping peaks.

As with in vivo voltammetry, a variety of electrochemical techniques have been used for the stripping step. Because of its simplicity, linear sweep voltammetry has enjoyed widespread use; however, the detection limit of this technique is limited by charging current. Differential pulse has become popular because it discriminates against the charging current to provide considerably lower detection limits.

Two types of mercury electrodes are commonly used for ASV — the hanging mercury drop electrode (HMDE) and the mercury film electrode (MFE). The HMDE consists of a drop of mercury usually suspended from a mercury thread in a glass capillary. A typical drop size is 0.5 mm in diameter. A suitable mercury drop can be extruded through a narrow glass capillary by means of a commercially available micrometer syringe or by the static mercury drop electrode, which is a solenoid controlled mercury capillary. A new drop is formed by simply dislodging the old one and extruding more mercury. Such devices are capable of producing drops of uniform area to within ~5%.

The main advantages of the HMDE are the very low residual current, the excellent negative potential range which pure mercury exhibits, and the ease and reliability with which reproducible drops can be extruded from a capillary (although clogging of the capillary in some solvents is occasionally an annoying problem). By comparison with the MFE described below, disadvantages of the HMDE include a limitation to rather slow stirring rates during the deposition step and restrictions on rotating the electrode without dislodging the drop; a low surface area/volume ratio, which necessitates longer deposition times; and broader stripping peaks, which can lead to loss of resolution in multielement analysis. In many analyses, these disadvantages are unimportant, and the HMDE serves as an excellent electrode.

The MFE consists of a thin mercury coating (1–1000 nm) on an electrically conductive substrate which provides physical stability to the liquid film. The usual procedure for preparing the MFE is by electrochemical reduction of Hg^{2+} to Hg on a conductive substrate such as glassy carbon. Procedures using both preformed films and in situ-deposited films have been used extensively. Preformed films are typically made by deposition of mercury at −0.4 to −1.0 V vs SCE from an acidic Hg^{2+} solution. The electrode is then removed from the plating solution and used for sample analysis. In situ-deposited films are formed simultaneously with deposition of sample metals by adding a small amount of Hg^{2+} to all standard and sample solutions. The mercury film is removed after each analysis either by wiping it off or by anodically stripping it to soluble Hg^{2+}, in the absence of anions which may cause precipitation.

The MFE exhibits certain advantageous characteristics which provide substantial impetus for its usage. An advantage of the MFE results from the greater electrode surface area/volume ratio of a film as compared to the spherical HMDE. The larger surface area enables more sample to be concentrated into a given amount of mercury during a specified deposition time. This means greater analytical sensitivity for the

MFE compared to the HMDE (or a shorter deposition time for the MFE for equal sensitivity with the HMDE). The MFE is also compatible with stirring rates which would dislodge a HMDE. The ability to use greater stirring rates or to rapidly rotate the electrode improves mass transport during the concentration step, which further reduces the time required to deposit a given quantity of sample. Another advantage of the MFE results from its thinness compared to the HMDE. In a thin film, the electrodeposited sample is constrained near the interphase to which it must diffuse for oxidation during the stripping step. By comparison, metals deposited into a HMDE can diffuse to the center of the spherical electrode, which lengthens the return trip to the interphase for oxidation. The practical consequence of this is much sharper stripping peaks for a MFE than a HMDE. This minimizes the overlap of stripping peaks in multielement analysis, reducing the uncertainties in baseline estimation for the measurement of peak current.

Since different metals strip from mercury electrodes at characteristic peak potentials, several metal ions can be determined simultaneously. Metal ions which have been determined by ASV at a mercury electrode are Bi^{3+}, Cd^{2+}, Cu^{2+}, Ga^{3+}, Ge^{4+}, In^{3+}, Ni^{2+}, Pb^{2+}, Sb^{3+}, Sn^{2+}, Tl^{+}, and Zn^{2+}. Solid electrodes such as graphite enable Hg^{2+}, Au^{3+}, Ag^{+}, and Pt^{6+} to be determined by ASV. In this case, the metal is preconcentrated on the surface of the electrode as a metallic film, which is then stripped off by the positive potential scan.

The determination of lead in blood is the most widespread clinical use of ASV [323]. The technique is attractive because it is rapid, simple and reproducible [324]. A recent advance is to couple ASV to flow injection analysis in order to automate the process so that smaller samples and shorter analysis time can be achieved [325]. Lead is also routinely determined in bonemeal meant for human consumption by ASV [326]. Both lead and cadmium are determined in agricultural crops by ASV [327, 328].

4 References

1. Arnold, M. A., Meyerhoff, M. E.: Anal. Chem. *56*, 20R (1984)
2. Arnold, M. A., Solsky, R. L.: ibid. *58*, 84R (1986)
3. Durst, R. A. (ed.): Ion-Selective Electrodes, Washington, D.C.: National Bureau of Standards 1969
4. Freiser, H. (ed.): Ion-Selective Electrodes in Analytical Chemistry, vols. I, II, New York: Plenum 1978, 1980
5. Covington, A. K. (ed.): Ion-Selective Electrodes Methodology, vols. I, II, Boca Raton, Florida: CRC Press 1979
6. Moody, G. J., Thomas, J. D. R.: Selective Ion Sensitive Electrodes, Watford England: Merrow Publishing Co. Ltd. 1971
7. Koryta, J.: Ion-Selective Electrodes, Cambridge: Cambridge University Press 1975
8. Bailey, P. L.: Analysis with Ion-Selective Electrodes, 2nd Ed., London: Heyden 1980
9. Morf, W. E.: The Principles of Ion-Selective Electrodes and of Membrane Transport, New York: Am. Elsevier 1981
10. Simon, W., Ammann, D., Bussmann, W., Meier, P. C.: Ion-Selective Electrodes in Biology and Medicine, in: Frontiers of Chemistry (Laidler, K. J., ed.) New York: Pergamon Press, 1982, p. 217
11. Meyerhoff, M. E., Opdycke, W. N.: Ion-Selective Electrodes, in: Advances in Clinical Chemistry, Vol. 25 (Spiegel, H. E., ed.) New York: Academic Press 1986, p. 1
12. Solsky, R. L.: Ion-Selective Electrodes in Biomedical Analysis, in: CRC Critical Reviews in Analytical Chemistry *14*, 1 (1982)

13. Tonelli, D.; Budini, R., Gattavecchia, E., Girotti, S.: Anal. Biochem. *111*, 189 (1981)
14. Parris, N., Foglia, T. A.: J. Agric. Food Chem. *31*, 887 (1983)
15. Cooke, R. J., Jensen, R. L., Clin. Chem. (Winston-Salem, N. C.): *29*, 867 (1983); *29*, 1563 (1983)
16. Corcoran, C. A., Rechnitz, G. A.: Tr. Biotech. *3*, 92 (1985)
17. Arnold, M. A.: Am. Lab. *15(6)*, 34 (1983)
18. Rechnitz, G. A.: Science *214*, 287 (1981)
19. Kobos, R. K.: Potentiometric Enzyme Methods, in: Ion-Selective Electrodes in Analytical Chemistry, Vol. 2 (Freiser, H., ed.) New York: Plenum Press, 1980, p. 1
20. Rechnitz, G. A., Kobos, R. K., Riechel, S. J., Gebauer, C. R.: Anal. Chim. Acta *94*, 357 (1977)
21. Corcoran, C. A., Kobos, R. K.: Anal. Lett. *16(B16)*, 1291 (1983)
22. Kobos, R. K., Rechnitz, G. A.: ibid. *10*, 751 (1977)
23. Rechnitz, G. A., Riechel, T. L., Kobos, R. K., Meyerhoff, M. E.: Science *199*, 440 (1978)
24. Riechel, T. L., Rechnitz, G. A.: J. Membr. Sci. *4*, 243 (1978)
25. Kobos, R. K., Pyon, H. Y.: Biotechnol. Bioeng. *23*, 627 (1981)
26. DiPaolantonio, C. L., Rechnitz, G. A.: Anal. Chim. Acta *141*, 1 (1982)
27. Walters, R. R., Moriarty, B. E., Buck, R. P.: Anal. Chem. *52*, 1680 (1980)
28. DiPaolantonio, C. L., Arnold, M. A., Rechnitz, G. A.: Anal. Chim. Acta *128*, 121 (1981)
29. Kobos, R. K., Rice, D. J., Flournoy, D. S.: Anal. Chem. *51*, 1122 (1979)
30. Kawashima, T., Tomida, K., Tominaga, N., Kobayashi, T., Onishi, H.: Chem. Lett. (Chem. Soc. Japan) *5*, 653 (1984)
31. Hikuma, M., Obana, H., Yasuda, T., Karube, I., Suzuki, S.: Anal. Chim. Acta *116*, 61 (1980)
32. DiPaolantonio, C. L., Rechnitz, G. A.: ibid. *148*, 1 (1983)
33. Jensen, M. A., Rechnitz, G. A.: ibid. *101*, 125 (1978)
34. Grobler, S. R., Rechnitz, G. A.: Talanta *27*, 283 (1980)
35. Ma, Y. L., Rechnitz, G. A.: Anal. Lett. *18*, 1635 (1985)
36. Smit, N., Rechnitz, G. A.: Biotech. Lett. *6*, 209 (1984)
37. Arnold, M. A., Rechnitz, G. A.: Anal. Chem. *53*, 515 (1981)
38. Arnold, M. A., Rechnitz, G. A.: ibid. *53*, 1837 (1981)
39. Kuriyama, S., Rechnitz, G. A.: Anal. Chim. Acta *131*, 91 (1981)
40. Rechnitz, G. A., Arnold, M. A., Meyerhoff, M. E.: Nature (London) *278*, 466 (1979)
41. Meyerhoff, M. E.: Anal. Lett. *13*, 1345 (1980)
42. Arnold, M. A., Rechnitz, G. A.: Anal. Chem. *54*, 777 (1982)
43. Grobler, S. R., Basson, N., Van Wyk, C. W.: Talanta *29*, 49 (1982)
44. Guilbault, G. G., Montalvo, J. G.: J. Am. Chem. Soc. *91*, 2164 (1969)
45. Guilbault, G. G., Montalvo, J. G.: ibid. *92*, 2533 (1970)
46. Guilbault, G. G., Montalvo, J. G.: Anal. Lett. *2*, 283 (1969)
47. Tran-Minh, C., Selegny, E., Broun, G.: C. R. Acad. Sci. Paris Ser. *C275*, 309 (1972)
48. Guilbault, G. G., Nagy, G.: Anal. Chem. *45*, 417 (1973)
49. Anfalt, T., Graneli, A., Jagner, D.: Anal. Lett. *6*, 969 (1973)
50. Guilbault, G. G., Tarp, M.: Anal. Chim. Acta *73*, 355 (1974)
51. Papastathopoulos, D. S., Rechnitz, G. A.: ibid. *79*, 17 (1975)
52. Mascini, M., Guilbault, G. G.: Anal. Chem. *49*, 795 (1977)
53. Guilbault, G. G., Shu, F. R.: ibid. *44*, 2161 (1972)
54. Tran-Minh, C., Broun, G.: C. R. Acad. Sci. Paris Ser. *D 276*, 2215 (1973)
55. Nilsson, H., Akerlund, A. C., Mosbach, K.: Biochim. Biophys. Acta *320*, 529 (1973)
56. Nagy, G., Von Storp, L. H., Guilbault, G. G.: Anal. Chim. Acta *66*, 443 (1973)
57. Rechnitz, G. A., Llenado, R. A.: Anal. Chem. *43*, 283 (1971)
58. Llenado, R. A., Rechnitz, G. A.: ibid. *43*, 1457 (1971)
59. Mascini, M., Liberti, A.: Anal. Chim. Acta *68*, 177 (1974)
60. Paperiello, G. J., Mukherji, A. K., Shearer, C. M.: Anal. Chem. *45*, 790 (1973)
61. Cullen, L. F., Rusling, J. F., Schleifer, A., Papariello, G. J.: ibid. *46*, 1955 (1974)
62. Guilbault, G. G., Hrabankova, E.: Anal. Lett. *3*, 53 (1970)
63. Guilbault, G. G., Hrabankova, E.: Anal. Chem. *42*, 1779 (1970)
64. Guilbault, G. G., Hrabankova, E.: Anal. Chim. Acta *56*, 285 (1971)
65. Berjonneau, A. M., Thomas, D., Broun, G.: Pathol. Biol. *22*, 497 (1974)
66. Calvot, C., Berjonneau, A. M., Gellf, G., Thomas, D.: FEBS Lett. *59*, 258 (1975)

67. Guilbault, G. G., Shu, F. R.: Anal. Chim. Acta *56*, 333 (1971)
68. Guilbault, G. G., Nagy, G.: Anal. Lett. *6*, 301 (1973)
69. Hsiung, C. P., Kuan, S. S., Guilbault, G. G.: Anal. Chim. Acta *90*, 45 (1977)
70. Wawro, R., Rechnitz, G. A.: J. Membr. Sci. *1*, 143 (1976)
71. White, W. C., Guilbault, G. G.: Anal. Chem. *50*, 1481 (1978)
72. Davies, P., Mosbach, K.: Biochim. Biophys. Acta *370*, 329 (1974)
73. Ahn, B. K., Wolfson, S. K., Yao, S. J.: Bioelectrochem. Bioenerg. *2*, 142 (1975)
74. Papastathopoulos, D. S., Rechnitz, G. A.: Anal. Chem. *48*, 862 (1976)
75. Rechnitz, G. A., Papastathopoulos, D. S., Saffran, M.: Fed. Proc. Fed. Am. Soc. Exp. Biol. *36*, 687 (1977)
76. Kawashima, T., Rechnitz, G. A.: Anal. Chim. Acta *83*, 9 (1976)
77. Meyerhoff, M., Rechnitz, G. A.: ibid. *85*, 277 (1976)
78. Tran-Minh, C., Guyonnet, R., Beaux, J.: C. R. Acad. Sci. Paris Ser. C, *286*, 115 (1978)
79. Durand, P., David, A., Thomas, D.: Biochim. Biophys. Acta *527*, 277 (1978)
80. Jensen, M. A., Rechnitz, G. A.: J. Membr. Sci. *5*, 117 (1979)
81. Shinbo, T., Sugiura, M., Kamo, N.: Anal. Chem. *51*, 100 (1979)
82. Tipton, K. F., McCrodden, J. M., Bardsley, M. E.: Biochem. Soc. Trans. *9*, 324 (1981)
83. Deng, I., Enke, C.: Anal. Chem. *52*, 1937 (1980)
84. Valle-Vega, P., Young, C. T., Swaisgood, H. E.: J. Food Sci. *45*, 1026 (1980)
85. Guilbault, G. G., Chen, S. P., Kuan, S. S.: Anal. Lett. *13*, 1607 (1980)
86. Arnold, M. A., Rechnitz, G. A.: Anal. Chem. *52*, 1170 (1980)
87. Nikolelis, D. P., Papastathopoulos, D. S., Hadjiioannou, T. P.: Anal. Chim. Acta *126*, 43 (1981)
88. Walters, R. R., Johnson, P. A., Buck, R. P.: Anal. Chem. *52*, 1684 (1980)
89. Kobos, R. K., Ramsey, T. A.: Anal. Chim. Acta *121*, 111 (1980)
90. Alexander, P. W., Jose, P. J.: ibid. *131*, 103 (1981)
91. Meyerhoff, M. E.: Anal. Lett. *13*, 1345 (1980)
92. Shinozuka, N., Hayano, S.: Nippon Kagaku Kaishi *10*, 1651 (1980); Anal. Abstr., 5J124
93. Hato, M., Suda, Y., Tsuda, K.: Nippon Kagaku Kaishi *7*, 1172 (1981); Chem. Abstr. *95*, 76309q
94. Kawashima, T., Arima, A., Hatakeyama, N., Tominaga, N., Ando, M.: Nippon Kagaku Kaishi *10*, 1542 (1980); Anal. Abstr., 4J97
95. Mohd, A. R., Riechel, T. L.: Anal. Lett. *17*, 835 (1984)
96. Al-Hitti, I. K., Moody, G. J., Thomas, J. D. R.: Analyst (London) *109*, 1205 (1984)
97. Ho, M. H., Wu, T. G.: Ann. N.Y. Acad. Sci. *434*, 523 (1984)
98. Yanushyavichyute, B. J., Panlyukonis, A. B., Kazlanskas, D. A.: Anal. Chem. (USSR) *38*, 498 (1983) (Russ.). Translation *38*, 392 (1983)
99. Seegopaul, P., Rechnitz, G. A.: Anal. Chem. *56*, 852 (1984)
100. Fonong, T., Rechnitz, G. A.: Anal. Chim. Acta *158*, 357 (1984)
101. Yasuda, K., Miyagi, H., Hamada, Y. Takata, Y.: Analyst (London) *109*, 61 (1984)
102. Pau, C. P., Rechnitz, G. A.: Anal. Chim. Acta *160*, 141 (1984)
103. Nikolelis, D. P., Hadjiioannou, T. P.: Anal. Let. *16*, 401 (1983)
104. Nikolelis, D. P., Hadjiioannou, T. P.: Anal. Chim. Acta *147*, 33 (1983)
105. Guilbault, G. G., Coulet, P. R.: ibid. *152*, 223 (1983)
106. Tran, N. D., Romette, J. L., Thomas, D.: Biotechnol. Bioeng. *25*, 329 (1983)
107. Havas, J., Guilbault, G. G.: Anal. Chem. *54*, 1991 (1982)
108. Vadgama, P. M., Alberti, K. G. M. M., Covington, A. K.: Anal. Chim. Acta *136*, 403 (1982)
109. Ianniello, R. M., Yacynych, A. M.: ibid. *146*, 249 (1983)
110. Ellard, P., Landet, A., Vire, J. C., Patriarche, G. J.: Analusis *10*, 182 (1982); (Fr.) Chem. Abstr. *97*, 3132h (1982)
111. Fischer, M., Landet, A., Vire, J. C., Patriarche, G. J., Dufrane, S. P.: J. Pharm. Belg. *37*, 202 (1982) (Fr.) Chem. Abstr. *97*, 178222q (1982)
112. Fung, K. W., Kuan, S. S., Sung, H. Y., Guilbault, G. G.: Anal. Chem. *51*, 2319 (1979)
113. Bergveld, P.: Bionsensors *2*, 15 (1986)
114. Janata, J.: Chemically Sensitive Field Effect Transistors, in: Solid State Chemical Sensors (Janata, J., Huber, R. J., eds.) New York: Academic Press 1985, p. 65
115. Ross, P.: Biotechniques *1*, 204 (1983)
116. Wohltjen, H.: Anal. Chem. *56*, 87A (1984)
117. Caras, S., Janata, J.: ibid. *52*, 1935 (1980)

118. Niyhara, Y., Matsu, F., Moriizumi, T.: Proc. Int. Meet. Chem. Sens., 1983, p. 501
119. Hanazato, Y., Shono, S.: Proc. Int. Meet. Chem. Sens., 1983, p. 513
120. Ammann, D.: Ion-Selective Microelectrodes, New York: Springer-Verlag 1986
121. Sykova, E., Hnik, P., Vyklicky, L. (eds.): Ion-Selective Microelectrodes and Their Use in Excitable Tissues, New York: Plenum Press 1981
122. Zeuthen, T. (ed.): The Application of Ion-Selective Microelectrodes, New York: Elsevier/North-Holland Biomedical Press 1981
123. Berman, H. J., Hebert, N. C. (eds.): Ion-Selective Microelectrodes, New York: Plenum Press 1974
124. Caldwell, P. C.: J. Physiol. (London) *126*, 169 (1954)
125. Hinke, J. A. M.: Nature (London) *184*, 1257 (1959)
126. Czaban, J. D., Rechnitz, G. A.: Anal. Chem. *45*, 471 (1973)
127. Steiner, R. A., Oehme, M., Ammann, D., Simon, W.: ibid. *51*, 351 (1979)
128. Lanter, F., Erne, D., Ammann, D., Simon, W.: ibid. *52*, 2400 (1980)
129. Ammann, D., Lanter, F., Steiner, R. A., Schulthess, P., Shijo, Y., Simon, W.: ibid. *53*, 2267 (1981)
130. Jaramillo, A., Lopez, S., Justice, J. B., Salamone, J. D., Neill, D. B.: Anal. Chim. Acta *146*, 149 (1983)
131. Walker, J. L.: Anal. Chem. *43*, 89A (1971)
132. Walker, J. L.: Single Cell Measurement with Ion-Selective Electrodes, in: Medical and Biological Applications of Electrochemical Devices (Koryta, J., ed.) New York, Wiley, 1980, p. 109
133. Khuri, R. N.: Proc. Int. Union Physiol. *9*, 301 (1971)
134. Zeuthen, T.: Curr. Top. Membr. Transport *13*, 31 (1980)
135. Kessler, M., Hajek, K., Simon, W.: Four-Barreled Microelectrode for the Measurement of Potassium, Sodium, and Calcium-Ion Activity, in: Ion and Enzyme Electrodes in Biology and Medicine (Kessler, M., Clark, Jr., L. C., Lubbers, D. W., Silver, I. A., Simon, W., eds.) Munich: Urban and Schwarzenberg, 1976, p. 136
136. Walker, J. L., Ladle, R. O.: Intracellular Chloride Activity in Heart Muscle, in: Ion and Enzyme Electrodes in Biology and Medicine (Kessler, M., Clark, Jr., L. C., Lubbers, D. W., Silver, I. A., Simon, W., eds.) Munich: Urban & Schwarzenberg, 1976, p. 351
137. Walker, J. L., Ladle, R. O.: Am. J. Physiol. *232*, E553 (1973)
138. Lee, C. O., Uhm, D. Y., Dresdner, K.: Science *209*, 699 (1980)
139. Owen, J. D., Brown, H. M., Pemberton, J. P.: Anal. Chim. Acta *90*, 241 (1977)
140. Kurkdjian, A. C., Barbier-Brygoo, H.: Anal. Biochem. *132*, 96 (1983)
141. Khuri, R. N.: Intracellular potassium in single cells of renal tubules, in: Ion and Enzyme Electrodes in Biology and Medicine (Kessler, M., Clark, Jr., L. C., Lubbers, D. W., Silver, I. W., Simon, W., eds.) Munich: Urban & Schwarzenberg, 1976, p. 364
142. Fujimoto, M., Kubota, T.: Jpn. J. Physiol. *26*, 631 (1976)
143. Christoffen, G. R. J., Simonsen, L.: Acta Physiol. Scand. *101*, 492 (1977)
144. Boron, W. F., DeWeer, P.: J. Gen. Physiol. *67*, 91 (1976)
145. Khuri, R. N., Hajjar, J. J., Agulian, S. K.: J. Appl. Physiol. *32*, 419 (1972)
146. Eaton, D. C., Russell, J. M., Brown, A. M.: J. Membr. Biol. *21*, 353 (1975)
147. Ellis, D.: J. Physiol. (London) *266*, 74P (1977)
148. Vaughn-Jones, R. D.: ibid. *272*, 32 P (1977)
149. Palmer, L. G., Civan, M. M.: J. Membr. Biol. *33*, 41 (1977)
150. Fogt, E. J., Eddy, A. R., Clemens, A. H., Fox, J., Heath, H., III: Clin. Chem. (Winston-Salem, N. C.) *26*, 1425 (1980)
151. Monroe, D.: Amer. Biotech. Lab. *4(6)*, 28 (1986)
152. Keating, M. Y., Rechnitz, G. A.: Anal. Chem. *56*, 801 (1984)
153. Keating, M. Y., Rechnitz, G. A.: Analyst *108*, 766 (1983)
154. Solsky, R. L., Rechnitz, G. A.: Anal. Chim. Acta *123*, 135 (1981)
155. Bush, D. L., Rechnitz, G. A.: Z. Anal. Chem. *333*, 491 (1986)
156. Aizawa, M., Kato, S., Suzuki, S.: J. Membr. Sci. *2*, 125 (1977)
157. Janata, J.: J. Am. Chem. Soc. *97*, 2914 (1975)
158. Yamamoto, N., Nagasawa, Y., Shuto, S., Sawai, M., Sudo, T., Tsubomura, H.: Chem. Lett. 245 (1978)

159. Yamamoto, N., Nagasawa, Y., Sawai, M., Sudo, T., Tsubomura, H.: J. Immunol. Mtds. *22*, 309 (1978)
160. Yamamoto, N., Nagasawa, Y., Shuto, S., Tsubomura, H., Sawai, M., Okumura, H.: Clin. Chem. *26*, 1569 (1980)
161. Janata, J., Huber, R. J. in: Ion-Selective Electrodes in Analytical Chemistry, Vol. 2 (Freiser, H., ed.) New York, Plenum Press, 1980, p. 107
162. Collins, S., Janata, J.: Anal. Chim. Acta *136*, 93 (1982)
163. Meyerhoff, M., Rechnitz, G. A.: Science *195*, 494 (1977)
164. Boitieux, J. L., Desmet, G., Thomas, D.: Clin. Chem. Acta *88*, 329 (1978)
165. Boitieux, J. L., Desmet, G., Thomas, D.: Clin. Chem. *25*, 318 (1979)
166. Boitieux, J. L., Lemay, D., Desmet, G., Thomas, D.: Clin. Chim. Acta *113*, 175 (1981)
167. Gebauer, C. R., Rechnitz, G. A.: Anal. Biochem. *103*, 280 (1980)
168. Meyerhoff, M. E., Rechnitz, G. A.: ibid. *95*, 483 (1979)
169. Alexander, P. W., Maltra, C.: Anal. Chem. *54*, 68 (1982)
170. Gebauer, C. R., Rechnitz, G. A.: Anal. Biochem. *124*, 338 (1982)
171. Brontman, S. B., Meyerhoff, M. E.: Anal. Chim. Acta *162*, 363 (1984)
172. Fonong, T., Rechnitz, G. A.: Anal. Chem. *56*, 2586 (1984)
173. Keating, M. Y., Rechnitz, G. A.: Anal. Lett. *18*, 1 (1985)
174. Shiba, K., Umezawa, Y., Watanabe, T., Ogawa, S., Fujihira, S.: Anal. Chem. *52*, 1610 (1980)
175. Shiba, K., Watanabe, T., Umezawa, Y., Fujiwara, S.: Chem. Lett. 155, (1980)
176. Umezawa, Y., Sofue, S., Takamoto, Y.: Talanta *31*, 375 (1984)
177. D'Orazio, P., Rechnitz, G. A.: Anal. Chem. *49*, 2083 (1977)
178. Shoup, R. E.: Recent Reports on Liquid Chromatography/Electrochemistry, West Lafayette: BAS Press 1982
179. Kissinger, P. T., Refshauge, D. J., Dreiling, R., Adams, R. N.: Anal. Lett. *6*, 465 (1973)
180. Fleet, B., Little, C. J.: J. Chromatogr. Sci. *12*, 747 (1974)
181. Blaedel, W. J., Schieffer, G. W.: J. Electroanal. Chem. *80*, 259 (1977)
182. Armentrout, D. N., McLean, J. D., Long, M. W.: Anal. Chem. *51*, 1039 (1979)
183. Koile, R. C., Johnson, D. C.: ibid. *51*, 741 (1979)
184. Messner, J. L., Engstrom, R. C.: ibid. *53*, 128 (1981)
185. Knecht, L. A., Guthrie, E. J., Jorgenson, J. W.: ibid. *56*, 479 (1984)
186. Roston, D. A., Shoup, R. E., Kissinger, P. T.: ibid. *54*, 1417A (1982)
187. Braun, R. C., J. Electroanal. Chem. *19*, 23 (1968)
188. Roston, D. A., Kissinger, P. T.: Anal. Chem. *54*, 429 (1982)
189. Lunte, C. E., Kissinger, P. T.: ibid. *56*, 658 (1984)
190. Lunte, C. E., Kissinger, P. T.: J. Liq. Chromatogr. *6*, 1863 (1983)
191. Lunte, C. E., Kissinger, P. T.: Anal. Chem. *55*, 1458 (1983)
192. Mayer, G. S., Shoup, R. E.: J. Chromatogr. *255*, 533 (1983)
193. Roston, D. A., Kissinger, P. T.: Anal. Chem. *53*, 1695 (1981)
194. Goto, M., Zou, G., Ishii, D.: J. Chromatogr. *268*, 157 (1983)
195. Lunte, C. E., Kissinger, P. T., Shoup, R. E.: Anal. Chem. *57*, 1541 (1985)
196. Kissinger, P. T., Bruntlett, C. S., Shoup, R. E.: Life Sci. *28*, 455 (1981)
197. Mefford, I. N.: J. Neurosci. Methods *3*, 207 (1981)
198. Kissinger, P. T.: Methods in Biog. Amine Res., (eds. Parvez, S., Nagatsu, T., Nagatsu, I.), Amsterdam: Elsevier 1983, pp. 75
199. Koch, D. D., Kissinger, P. T.: J. Chromatogr. *164*, 441 (1979)
200. Mefford, I. N., Barchas, J. D.: ibid. *181*, 187 (1980)
201. Anderson, G. M., Young, J. G., Cohen, D. J.: ibid. *228*, 155 (1982)
202. Seegal, R. F., Brosch, K. O., Bush, B.: ibid. *377*, 131 (1986)
203. Lunte, S. M., Kissinger, P. T.: Chem.-Biol. Interactions *47*, 195 (1983)
204. Miner, D. J., Kissinger, P. T.: Biochem. Pharmacol. *28*, 3285 (1979)
205. Hamilton, M., Kissinger, P. T.: Anal. Biochem. *125*, 143 (1982)
206. Radzik, D. M., Roston, D. A., Kissinger, P. T.: ibid. *131*, 458 (1983)
207. Shoup, R. E., Mayer, G. S.: Anal. Chem. *54*, 1164 (1982)
208. Kenyhercz, T. M., Kissinger, P. T.: J. Agr. Food Chem. *25*, 959 (1977)
209. Roston, D. A., Kissinger, P. T.: Anal. Chem. *53*, 1695 (1981)
210. Lunte, S. M., J. Chromatogr. *384*, 371 (1987)

211. Rice, J. R., Kissinger, P. T.: J. Anal. Toxicol. *3*, 64 (1979)
212. Radzik, D. M., Kissinger, P. T.: Anal. Biochem. *140*, 74 (1984)
213. Radzik, D. M., Brodbelt, J. S., Kissinger, P. T.: Anal. Chem. *56*, 2927 (1984)
214. Ward, C. J. P., Radzik, D. M., Kissinger, P. T.: J. Liq. Chromatogr. *8*, 677 (1985)
215. Armentrout, D. N., Cutie, S. S.: J. Chromatogr. Sci. *18*, 370 (1980)
216. Pachla, L. A., Kissinger, P. T.: Anal. Chem. *48*, 364 (1976)
217. Mason, W. D., Amick, E. M., Heft, W.: Anal. Lett. *13*, 817 (1980)
218. Pachla, L. A., Kissinger, P. T.: Clin. Chim. Acta *59*, 309 (1975)
219. Green, D. J., Perlman, R. L.: Clin. Chem. *26*, 796 (1980)
220. Lankelma, J., VanDerKleijn, E., Jansen, M. J. Th.: J. Chromatogr. *182*, 35 (1980)
221. Lunte, C. E., Kissinger, P. T.: Anal. Biochem. *129*, 377 (1983)
222. Lunte, C. E., Kissinger, P. T.: Anal. Chim. Acta *158*, 33 (1984)
223. Lunte, C. E., Kissinger, P. T.: Methods in Enzymol., Vol. 122 New York: Academic Press 1986, pp. 300
224. Kamata, K., Hagiwara, T., Takahaski, M., Uehara, S., Nakayama, K., Akiyama, K.: J. Chromatogr. *356*, 326 (1986)
225. Davis, G. C., Holland, K. L., Kissinger, P. T.: J. Liq. Chromatogr. *2*, 663 (1979)
226. Kok, W. Th., Brinkman, U. A. Th., Frei, R. W.: J. Chromatogr. *256*, 17 (1983)
227. Jacobs, W. A., Kissinger, P. T.: J. Liq. Chromatogr. *5*, 881 (1982)
228. Allison, L. A., Myer, G. S., Shoup, R. E.: Anal. Chem. *56*, 1089 (1984)
229. Carlsson, A., Lundstrom, K.: J. Chromatogr. *350*, 169 (1985)
230. Fleming, L. H., Reynolds, Jr., N. C.: ibid. *375*, 65 (1986)
231. Allison, L. A., Shoup, R. E.: Anal. Chem. *55*, 8 (1983)
232. Lunte, S. M., Kissinger, P. T.: J. Chromatogr. *317*, 579 (1984)
233. Lunte, S. M., Kissinger, P. T.: J. Liq. Chromatogr. *8*, 691 (1985)
234. Bratin, K., Kissinger, P. T., Bruntlett, C. S.: ibid. *4*, 1777 (1981)
235. Davenport, R. J., Johnson, D. C.: Anal. Chem. *46*, 1971 (1974)
236. Bratin, K., King, W. P., Kissinger, P. T., Rice, J. P.: ACS Symp. Series *136*, 57 (1980)
237. Vohra, S. K., Harrington, G. W.: Food Cosmet. Toxicol. *19*, 485 (1981)
238. Rappaport, S. M., Jin, Z. L., Xu, X. B.: J. Chromatogr. *240*, 145 (1982)
239. Shimada, K., Tanaka, M., Nambara, T.: Anal. Lett. *13*, 1129 (1980)
240. Brunt, K., Bruins, C. H. P., Doornbos, D. A.: Anal. Chim. Acta *125*, 85 (1981)
241. Hart, J. P., Sheares, M. J., McCarthy, P. T.: Analyst *110*, 1181 (1985)
242. Lund, W., Hannisdal, M., Greibrokk, T.: J. Chromatogr. *173*, 249 (1979)
243. Hanekamp, H. B., Voogt, W. H., Frei, R. W., Bos, P.: Anal. Chem. *53*, 1362 (1981)
244. Caudill, W. L., Ewing, A. G., Jones, S., Wightman, R. M.: ibid. *55*, 1877 (1983)
245. Samuelsson, R., O'Dea, J., Osteryoung, J.: ibid. *52*, 2215 (1980)
246. Wang, J., Dewald, H. D.: Anal. Chim. Acta *153*, 325 (1983)
247. Last, T. A.: Anal. Chem. *55*, 1509 (1983)
248. White, J. G., St. Claire III, R. L., Jorgenson, J. W.: ibid. *58*, 293 (1986)
249. Lunte, C. E., Wong, S., Chan, K. W., Ridgway, T. H., Heineman, W. R.: Anal. Chim. Acta *188*, 263 (1986)
250. Lunte, C. E., Ridgway, T. H., Heineman W. R.: Anal. Chem. *57*, 761 (1987)
251. Wilson, J. M., Slattery, J. T., Forte, A. J., Nelson, S. D.: J. Chromatogr. *227*, 453 (1982)
252. Meinsma, D. A., Radzik, D. M., Kissinger, P. T.: J. Liq. Chromatogr. *6*, 2311 (1983)
253. Smith, R. V., Humphrey, D. W.: Anal. Lett. *14*, 601 (1981)
254. Raffa, R. B., O'Neill, J. J., Tallarida, R. J.: J. Chromatogr. *238*, 515 (1982)
255. Svensson, J. O.: ibid. *375*, 174 (1986)
256. Hynning, P., Anderson, P., Bondesson, U., Boreus, L. O.: ibid. *375*, 207 (1986)
257. Lund, W., Hannisdal, M., Greibrokk, T.: ibid. *173*, 249 (1979)
258. Lund, W., Opheim, L.-N.: Anal. Chim. Acta *88*, 275 (1977)
259. Hanekamp, H. B., Voogt, W. H., Bos, P., Frei, R. W.: J. Liq. Chromatogr. *3*, 1205 (1980)
260. Suckow, R. F., Cooper, T. B.: J. Pharm. Sci. *70*, 257 (1981)
261. Brooks, M. A., Hackman, M. R., Mazzo, D. J.: J. Chromatogr. *210*, 531 (1981)
262. Wang, J., Bonakdar, M., Deshmukh, B. K.: ibid. *344*, 412 (1985)
263. Krause, W.: ibid. *181*, 67 (1980)
264. Lankelma, J., Poppe, H.: ibid. *149*, 587 (1978)

265. Rucki, R. J., Ross, A., Moros, S. A.: ibid. *190*, 359 (1980)
266. Lewis, E. C., Johnson, D. C.: Clin. Chem. *24*, 1711 (1978)
267. Greenberg, M. S., Mayer, W. J.: J. Chromatogr. *169*, 321 (1979)
268. Park, G. B., Koss, R. F., O'Neil, S. K., Palace, G. P., Edelson, J.: Anal. Chem. *53*, 604 (1981)
269. Magic, S. E.: J. Chromatogr. *129*, 73 (1976)
270. Oldfield, S., Berg, J. D., Stiles, H. J. Buckley, B. M.: ibid. *377*, 423 (1986)
271. Morris, J., Shoup, R. E.: Clin. Liq. Chromatogr. *1*, 97 (1984)
272. Drummer, O. H., Chritophidis, N., Horowitz, J. D., Louis, W. J.: J. Chromatogr. *374*, 251 (1986)
273. Lunte, C. E., Kissinger, P. T.: Anal. Biochem. *139*, 468 (1984)
274. Blank, C. L., Pike, R.: Life Sci. *18*, 859 (1976)
275. Messripur, M., Clark, J. B.: J. Neurochem. *38*, 1139 (1982)
276. Rahman, Md. K., Nagatsu, T., Kato, T.: Life Sci. *28*, 485 (1981)
277. Davis, G. C., Kissinger, P. T.: Anal. Chem. *51*, 1960 (1979)
278. Trocewicz, J., Oka, K., Nagatsu, T.: J. Chromatogr. *227*, 407 (1982)
279. Shoup, R. E., David, G. C., Kissinger, P. T.: Anal. Chem. *52*, 483 (1980)
280. Kiuchi, K., Kiuchi, K., Nagatsu, T., Togari, A., Kumagai, H.: J. Chromatogr. *357*, 191 (1986)
281. Davis, G. C., Holland, K. L., Kissinger, P. T.: J. Liq. Chromatogr. *2*, 663 (1979)
282. Lequeu, B., Guilland, J.-C., Klepping, J.: Anal. Biochem. *149*, 296 (1985)
283. Potter, P. E., Meek, J. L., Neff, N. H.: J. Neurochem. *41*, 188 (1983)
284. Asano, M., Miyauchi, R., Kato, T., Fujimori, K., Yamamoto, K.: J. Liq. Chromatogr. *9*, 199 (1986)
285. Dalgaard, L., Nordholm, L., Brimer, L.: J. Chromatogr. *265*, 183 (1983)
286. Kamada, S., Maeda, M., Tsuji, A., Umezawa, Y., Kurahashi, T.: ibid. *239*, 773 (1982)
287. Attiyat, A. S., Christian, G. S.: Analyst *105*, 154 (1980)
288. Blaedel, W. J., Boguslaski, R. C.: Anal. Chem. *50*, 1026 (1978)
289. Wehmeyer, K. R., Halsall, H. B., Heineman, W. R., Volle, C. P., Chen, I.: ibid. *58*, 135 (1986)
290. Wehmeyer, K. R., Doyle, M. J., Wright, D. S., Eggers, H. M., Halsall, H. B., Heineman, W. R.: J. Liq. Chromatogr. *6*, 2141 (1983)
291. Doyle, M. J., Halsall, H. B., Heineman, W. R.: Anal. Chem. *56*, 2355 (1984)
292. Wehmeyer, K. R., Halsall, H. B., Heineman, W. R.: Clin. Chem. *31*, 1546 (1985)
293. Aizawa, M., Morioka, A., Suzuki, S.: Anal. Chim. Acta *115*, 61 (1980)
294. Itagaki, H., Hakoda, Y., Suzuki, Y., Haga, M.: Chem. Pharm. Bull. *31*, 1283 (1983)
295. Karube, I., Matsunaga, T., Suzuki, Y., Asano, T., Itah, S.: J. Biotech. *1*, 279 (1984)
296. Ikariyama, Y., Suzuki, S., Aizawa, M.: Anal. Chem. *54*, 1126 (1982)
297. de Alwis, W. U., Wilson, G. S.: ibid. *57*, 2754 (1985)
298. Eggers, H. M., Halsall, H. B., Heineman, W. R.: Clin. Chem. *28*, 1848 (1982)
299. Wright, D. S., Halsall, H. B., Heineman, W. R.: Anal. Chem.
300. Ngo, T. T., Bovaird, J. H., Lenhoff, H. M.: Appl. Biochem. Biotech. *11*, 63 (1985)
301. Kissinger, P. T., Hart, J. B., Adams, R. N.: Brain Res. *55*, 209 (1973)
302. Marsden, C. A. (ed.): Measurement of Neurotransmitter Release *in vivo*, IBRO Handbook Series, Vol. 6, Chickester, England: Wiley-Interscience 1984
303. Adam, R. N., Marsden, C. A.: Handbook in Psychopharmacology, Vol. 15 (Iverson, L. L., Iverson, S. D., Snyder, S. H., eds.) New York: Plenum Press, 1982, pp. 1–74
304. Schenk, J. O., Miller, E., Rice, M. E., Adams, R. N.: Brain Res. *277*, 1 (1983)
305. Lane, R. F., Hubbard, A. T., Blaha, C. D.: Anal. Chem. *48*, 1287 (1976)
306. Kennett, G. A., Joseph, M. H.: Brain Res. *236*, 305 (1982)
307. Kuhr, W. G., Ewing, A. G., Caudill, W. L., Wightman, R. M.: J. Neurochem. *43*, 560 (1984)
308. Brazell, M. P., Marsden, C. A.: Br. J. Pharmacol. *75*, 539 (1982)
309. Buda, M., DeSimoni, M. G., Gonon, F., Pujol, J.-F.: Brain Res. *273*, 197 (1983)
310. McCreery, R. I., Dreiling, R., Adams, R. N.: ibid. *73*, 23 (1974)
311. O'Neill, R. D., Grunewald, R. A., Fillenz, M., Alberg, W. J.: Neuroscience *7*, 1945 (1982)
312. Conti, J., Strope, E., Adams, R. N., Marsden, C. A.: Life Sci. *23*, 2705 (1978)
313. Mefford, I. N., Oke, A. F., Adam, R. N.: Brain Res. *212*, 223 (1981)
314. Crespi, F., Sharp, T., Maidment, N., Marsden, C. A.: Neurosci. Lett. *43*, 203 (1983)
315. Gonon, F. G., Fombarlet, C. M., Buda, M. J., Pujol, J. F.: Anal. Chem. *53*, 1386 (1981)
316. Falat, L., Cheng, H.-Y.: ibid. *54*, 2108 (1982)

317. Yamamoto, B. K., Lane, R. F., Freed, C. R.: Life Sci. *30*, 2155 (1982)
318. Gerhardt, G. A., Oke, A. F., Nagy, G., Adams, R. N.: Brain Res. *290*, 390 (1983)
319. Brazell, M. P., Marsden, C. A.: ibid. *249*, 167 (1982)
320. Nagy, G., Rice, M. E., Adams, R. N.: Life Sci. *31*, 2611 (1982)
321. Ewing, A. G., Dayton, M. A.: Wightman, R. M.: Anal. Chem. *53*, 1842 (1981)
322. Ewing, A. G., Wightman, R. M., Dayton, M. A.: Brain Res. *249*, 361 (1982)
323. Blanke, R. V., Decker, W. J.: Textbook of Clinical Chemistry, (eds. Tietz, N. W.), Philadelphia: W. B. Saunders 1986, p. 1712
324. Searle, B., Chan, W., Davidow, B.: Clin. Chem. *19*, 76 (1973)
325. Wise, J. A., Heineman, W. R.: Anal. Chim. Acta *172*, 1 (1985)
326. Satzger, R. D., Roy, W. K., Fricke, F. L.: J. Assoc. Off. Anal. Chem. *66*, 985 (1983)
327. Satzger, R. D., Clow, C. S., Bonnin, E., Fricke, F. L.: ibid. *65*, 987 (1982)
328. Satzger, R. D., Bonnin, R., Fricke, F. L.: ibid. *67*, 1138 (1984)

Medical Applications of Electrochemical Sensors and Techniques

Gary S. Calabrese and Kathleen M. O'Connell

Applied Research Department, Instrumentation Laboratory, 113 Hartwell Avenue, Lexington, Massachusetts, U.S.A.

Table of Contents

Four major areas of electrochemistry related to medical diagnostics have been reviewed. Blood pH and gas measurements as well as ISE's represent relatively mature areas which enjoy widespread commercialization. New approaches should yield devices which have superior performance and which are less expensive to produce. Enzyme electrodes and electrochemical immunoassay are still largely experimental, but the intense level of current research effort coupled with some interesting recent developments should lead to commercial success over the next decade.

1 Introduction

The needs of modern medicine require sophisticated measuring devices for a wide range of chemical species in very complex matricies [1]. Due to fundamental breakthroughs in the 1950's and 1960's, electrochemical methods have proven to be extremely valuable in meeting such needs. This past success, coupled with the current trend in the United States and western Europe towards cutting medical costs while still maintaining quality of care is largely responsible for the present fast pace of research and development in the area [1-4].

It is the purpose here to briefly review the state of the art of the most important electrochemical methods for medical applications, and report on the status and viability of currently emerging research. To accomplish this, electrochemical methods have been divided into four basic categories. The first two categories (Sect. 2 and 3) represent the relatively mature contribution of electrochemistry to medical diagnostics. Sections four and five deal largely with developments in electrochemistry which have not yet achieved commercialization, but which have the greatest likelihood of future success. There are, of course, some minor areas of research which have been intentionally omitted because of space limitations. Much of this work can be found in the references provided in the text.

2 Blood pH, pCO_2 and pO_2 Measurements

2.1 Introduction

Along with the measurement of pH, the measurement of the partial pressures of oxygen (pO_2) and carbon dioxide (pCO_2) in blood is an essential means for discovering the nature and origin of disturbances to the body's acid-base equilibrium. At present such measurements [5] represent one of the most commercially important applications of electrochemistry to medical diagnostics, and so it is not surprising that much work has been reported in this area over the years. However, despite the appearance of a number of different electrochemical and non-electrochemical techniques for each of the three analytes, the same fundamental technology [6,7] has been in continuous use for nearly 30 years. Some of the most interesting developments in the last decade or so have been in the area of new fabrication techniques and electrode miniaturization for in vivo applications, though commercialization of such systems is not yet widespread.

2.2 pH Measurements

In commercial existence since 1936, the glass pH electrode has been the workhorse of blood pH electrodes since the first three channel (pH, pCO_2 and pO_2) blood gas system was introduced in 1958. The subject of numerous reviews [8-12], the basic principle of operation which is discussed in more detail in Sect. 3 relies upon the potential developed across a glass membrane separating two electrolytes which arises from the cation exchange process of ions within the glass. The composition of one of the solutions remains constant, so that changes in potential can be used to determine changes in pH of the other solution (analyte phase). The time proven, although somewhat cumbersome, approach to accomplishing this for virtually all types of pH measurements relies on an "inner" reference electrode (most often Ag/AgCl) and an additional external reference electrode (often Ag/AgCl or saturated calomel) which is in direct contact with the analyte phase. Clever electrode designs have allowed for remarkably compact systems which can measure pH of very small volume samples. In addition, electrodes for in vivo use have been known for some time [13], although these systems have been historically plagued by poor precision (+0.05 pH units) due to excessive drift.

The present state of the art in blood pH measurements allows for rapid (1 minute) determination of pH between 6.4 and 8.0 to within at least 0.005 units for whole blood sample volumes <100 microliters. The temperature of the electrodes and sample is generally controlled to within 0.1 °C for this level of precision and frequent calibration is carried out (in some cases a "one point" calibration for each sample). The electrodes require (both the glass and external reference) some maintenance due to protein fouling, however this procedure is largely automated. The useful life of an electrode is one year or less and the cost is well over $ 100 (U.S.) each. New technologies, both electrochemical and non-electrochemical, must compete with this attractive performance and provide for lower operating costs in order to be successful.

By far, the most potentially significant advances in pH electrode technology over the last decade have come about because of the concurrent revolution in silicon semiconductor microfabrication technology. Known since 1970 [14,15], the chemically sensitive field-effect transistor (CHEMFET) is the most important embodiment of this. Figure 1 shows a schematic of the first CHEMFET reported, a pH sensitive device. Reversible hydration of the SiO_2 region at different values of pH gives changes in charge density in this region thus altering the field in the channel. These changes affect the conductivity between the source and drain which can be monitored by measuring changes in drain current at constant gate (V_g) and drain (V_D) voltages or by measuring changes in V_D at constant I_D and V_g. Signal amplification is effected since relatively small changes in interfacial potential at the liquid/solid interface can yield large changes in either V_D or I_D.

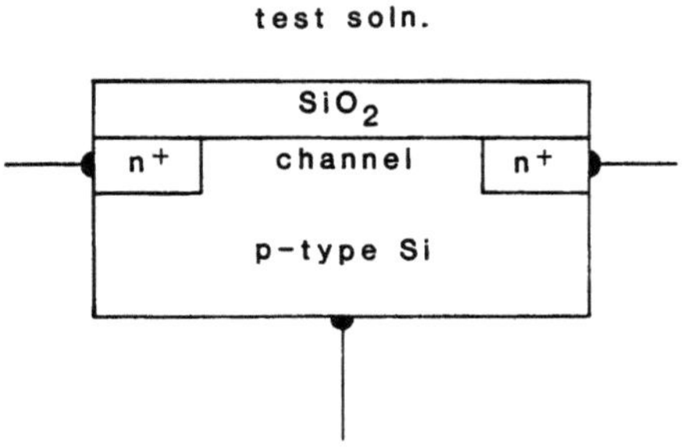

Fig. 1. Schematic diagram of the first reported pH responsive CHEMFET

In the early days of CHEMFET development, the expectations for successful application to a variety of biomedically important sensing applications were high. This was in part due to the fact that CHEMFET's are easily miniaturized (<2 mm^2 surface area) and so are obvious candidates for in vivo applications. This enthusiasm has largely been tempered by the reality that although field-effect tran-

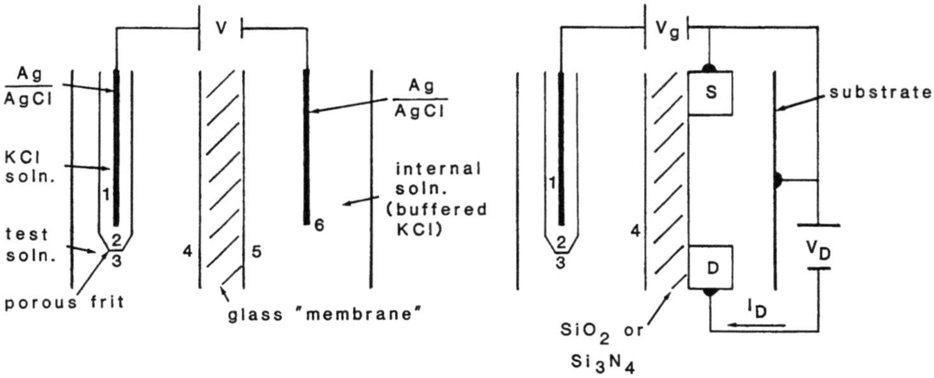

Fig. 2. Comparison of the typical configuration used for pH measurements (left) with a CHEMFET system (right)

sistors for electronics applications [16] may now be mass produced at low cost, the subsequent processing steps required to make these devices CHEMFET's which are selective for a given analyte and function with reasonable stability in harsh media such as blood or urine are not easily automated. This is especially true in cases where complex sensing structures such as enzyme-containing membranes are used.

Figure 2 graphically summarizes the difference between CHEMFET's and the traditional configuration used to make measurements of pH as well as other ions. A total of six liquid/solid interfaces are employed to measure changes in the potential difference across the single interface of interest, #4. The right-hand portion of the figure shows how a CHEMFET simplifies the situation, since two liquid/solid junctions have been replaced with a solid-state device. The necessary well-defined energetics associated with this replaced "inner" liquid/solid junction reference system are now determined by solid-state junctions which comprise the heart of a field-effect transistor. Modern semiconductor fabrication technology permits an excellent degree of control over the energetics of these solid-state junctions, thus allowing for the enormous success of field-effect transistors. In addition to the obvious advantages associated with having fewer liquid/solid junctions, the CHEMFET structure enables amplification virtually right at the site of signal generation. Thus noise problems associated with carrying the high impedance signals occuring from changes at junction #4 through wires over practical distances to where they can be amplified are minimized. This is especially important for in vivo applications since the diameter of shielded leads necessary with conventional electrodes can place limitations on how these devices can be implanted within the body.

Since the initial report of the pH responsive CHEMFET in 1970, CHEMFET's for other species such as Ca^{2+}, Na^{+}, K^{+} and penicillin have been described. In addition, some of these devices have been tested for in vivo or on-line continuous whole blood monitoring. While problems associated with mass production of the more complex CHEMFET's such as those employing enzymes (for example, with the penicillin [17] CHEMFET) have not yet been fully solved, the technology for mass production of the relatively simple pH CHEMFET is apparently now available [18–21], and problems noted with early devices attributable to irreversible SiO_2 changes and changes in the Si/SiO_2 interface have been largely overcome by using other gate materials such as Si_3N_4.

In addition to CHEMFET's, other types of chemically sensitive electronic devices are receiving attention, most notably, devices constructed on ceramic substrates entirely by inexpensive screen-printing procedures. Some early work [22)] in this area has employed an H^+ ionophore (tridodecylamine) in a polyvinyl chloride (PVC) support matrix as the active sensing layer. An energetically stable and reproducible contact is made to a silver conducting layer by using an intervening carbon layer. The early results are promising, but the precision demonstrated by these devices will have to be improved by roughly an order of magnitude to be competitive with existing technology.

Despite the advances in CHEMFET's and other chemically sensitive electronic devices, they have not yet achieved commercial success. Assuming the performance (precision, accuracy, response time, thermal sensitivity, durability, etc.) of these devices can match or exceed that of conventional pH electrodes, the only issue concerning their viability as alternatives is cost. With the apparent successes in automation of the entire CHEMFET process for pH devices it seems likely that some degree of commercialization will be achieved if attractive preliminary performance claims associated with some recently reported CHEMFET devices are corroborated.

2.3 pCO_2 Measurements

The pCO_2 electrode was first described in 1957 by Stow [23)] and later improved to its presently used form by Severinghaus [24)]. The basic principle of operation relies on equilibration of CO_2 with an aqueous solution. The change in pH in the aqueous solution associated with the equilibration due to carbonic acid formation (H_2CO_3) is measured and varies with log $[pCO_2]$ [25)]. It should be pointed out here that these measurements (and likewise for the pO_2 electrode described below) give CO_2 tension *not* concentration. To obtain concentration, Henry's Law of gas solubility must be applied. However, for most medical and biological applications the knowledge of the gas tension is sufficient.

Figure 3 is a schematic representation of a typical CO_2 electrode. A KCl/HCO_3^- containing electrolyte solution is trapped within a nylon mesh spacer layer whose pH is monitored by a contacting conventional glass pH electrode. A CO_2 permeable membrane isolates the electrolyte layer from the analyte phase. Currently available

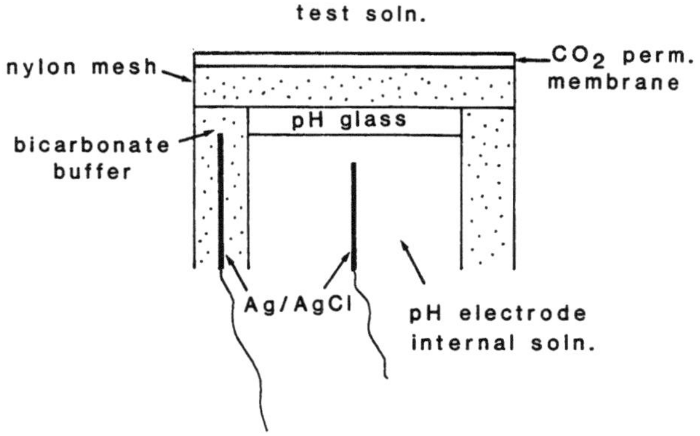

Fig. 3. Schematic diagram of a pCO_2 electrode

systems for blood pCO_2 can measure whole blood pCO_2 from 10 to 260 mm Hg in 0.1 mm steps. Sample volumes of 100 microliters or less are required and results are obtained in approximately one minute. Like blood gas pH electrodes, the CO_2 electrode needs regular maintenance and has a normal life of one year and costs well over $100 (U.S.).

One obvious approach to advancing pCO_2 technology is to replace the conventional pH electrode substructure with a CHEMFET. A much more unique approach [26] recently reported actually carries out a titration of carbonic acid from CO_2 using electrogenerated OH^- in the vicinity of a pH CHEMFET. The CHEMFET is used to measure relative changes in pH during the course of the titration, thus yielding measurement of CO_2 concentration without the apparent need for frequent calibration. Many of the operating characteristics of such an approach are attractive, but the demonstrated precision (10%) will have to be improved for biomedical applications.

Other important alternate electrochemical methods under study for pCO_2 rely on measuring current associated with the direct reduction of CO_2. The electrochemistry of CO_2 in both aqueous and non-aqueous media has been documented for some time [27-29], but interferences from more easily reduced species such as O_2 as well as many commonly used inhalation anesthetics [30] have made the direct amperometric approach difficult to implement. One recently described attempt [31,32] to circumvent some of these interference problems employs a two cathode configuration in which one electrode is used to "scrub" the sample of O_2 by exhaustive reduction prior to CO_2 amperometry at the second electrode. The response time and sensitivity of the approach may prove to be adequate for blood gas applications, but the issue of interfering anesthetics must be addressed more thoroughly in order to make the technique a truly viable alternative to the presently used indirect potentiometric electrode.

The ability of differential pulse polarography to resolve multicomponent systems and evaluate concentrations with excellent sensitivity has made this technique an attractive candidate for simultaneous measurement of CO_2, O_2, and some inhalation anesthetics [33]. Though very preliminary, the results appear promising and will likely lead to more intensive investigation of the approach.

2.4 pO_2 Measurements

Amperometry at inert metal cathodes is the most important approach to pO_2 measurements known today [34,35]. The subject of experimental investigations since 1945 [36], the amperometric approach was made commercially successful by the pioneering work of Clark in 1956 [37]. The so-called Clark sensor employs a platinum cathode and a silver-based anode which also functions as a reference electrode, Fig. 4. Both electrodes are in contact with a thin aqueous electrolyte layer which is covered by an O_2 permeable membrane. The cathode is poised at a potential with respect to the reference anode which is negative enough to effect reduction of oxygen to H_2O and/or H_2O_2, but not too negative such that reduction of H^+ occurs. The current is measured at some given time after sample introduction and is directly proportional to pO_2.

An enormous number of modifications of the basic Clark design have been reported over the years [38], most attempting to alter the speed, sensitivity and flow dependence of the measurement to suit a particular application. This is normally accomplished by

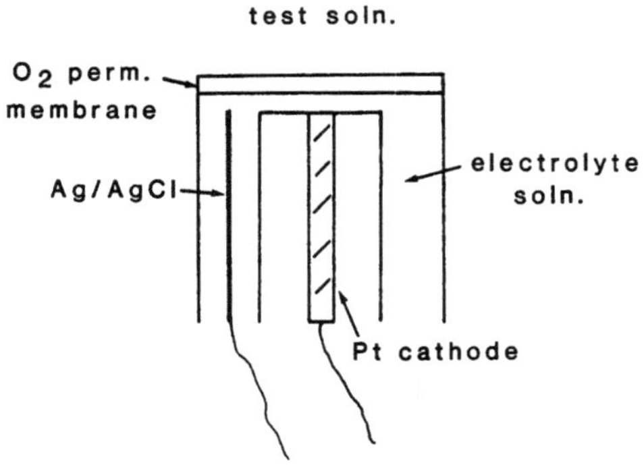

Fig. 4. Schematic diagram of a pO_2 electrode

changing the cathode area, and thickness of both the membrane and electrolyte layers. For blood gas applications, modern sensors are designed such that they can measure pO_2 from 0.0 to 800.0 mm Hg within one minute on whole blood samples < 100 microliters in volume. Like pH and pCO_2 measurements, the temperature must be controlled to within 0.1 °C to obtain the ± 0.1 mm Hg precision, calibration and maintenance are necessary, and sensor replacement is usually yearly at a cost comparable to pH and pCO_2 electrodes. Driven primarily by the perceived need for continuous oxygen monitoring, there have appeared a large number of reports of catheter pO_2 sensors [39], some of which appear to have good performance characteristics. The latest important advances in pO_2 sensors relate to the development of fabrication procedures to make them disposable in order to partially circumvent the issue of device durability. Use of silicon microfabrication procedures has been reported [40-42], but control of the deposition process for the liquid electrolyte layer and O_2 permeable membrane to make devices with reproducible characteristics was not clearly demonstrated.

More recently, the use of screen-printing procedures for pO_2 sensor fabrication has been reported [43,44]. Unlike the application of such procedures to pH sensors, there is a potential limitation in the case of pO_2 sensors since electrode sizes cannot be controlled much below 100 microns. This is important because cathode diameters of $\leqq 25$ microns are desirable since they minimize flow dependency, overcome the non-linearity problem which occurs from O_2 depletion with large electrodes, and lead to higher current densities due to radial diffusion components. In one of the reports [43], however, the O_2 depletion problem was avoided by using a voltage pulse technique to probe O_2 concentration and the claim was made that the sensors described are useful for single-use whole blood measurements, though detailed performance data were not provided. Pulse techniques for pO_2 measurements have been studied for some time and are generally conceded to suffer from difficulties which make them unattractive [45]. Thus, the apparent recent success [43] of the technique is significant and will probably revive interest in it.

3 Ion-Selective Electrodes

3.1 Introduction

The worldwide prevalence of ion-selective electrodes (ISE's) in clinical analyzers is a consequence of a major research effort which spanned more than two decades. The commercial success of ISE's for the determination of ionic species in body fluids continues to supplant the flame photometer in the clinical laboratory. Many interesting reviews have appeared in the last few years on ISE's in general [46–53] as well as their application in clinical analysis [54–61].

An ion-selective electrode contains a semipermeable membrane in contact with a reference solution on one side and a sample solution on the other side. The membrane will be permeable to either cations or anions and the transport of counter ions will be restricted by the membrane, and thus a separation of charge occurs at the interface. This is the Donnan potential (Fig. 5a) and contains the analytically useful information. A concentration gradient will promote diffusion of ions within the membrane. If the ionic mobilities vary greatly, a charge separation occurs (Fig. 5b) giving rise to what is called a diffusion potential.

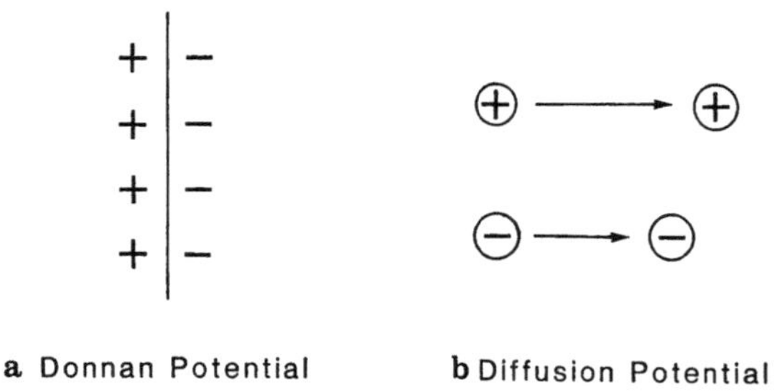

Fig. 5. Charge separation **a** at an interface with restricted transport and **b** with freely diffusing species

A single electrochemical potential, such as the ion-selective membrane potential, ΔE_M, cannot be measured directly. Rather, potentiometry involves the measurement of potential differences. Experimentally, an electrochemical cell comprising an indicating electrode (ISE) and a reference electrode is required. The energetics of a typical cell are shown in Fig. 6. The reference electrode and the ISE are in electrical contact via the sample solution. The reference electrode shown here has a liquid junction indicated by a double line which allows electrical contact but minimizes the leaching of the reference electrode filling solution into the sample. Commonly used junctions consist of a ceramic or glass frit, porous vycor glass, or cellophane. A junction diffusion potential, E_j, develops across this interface, but is minimized by using an electrolyte (KCl) whose components have similar ionic mobilities.

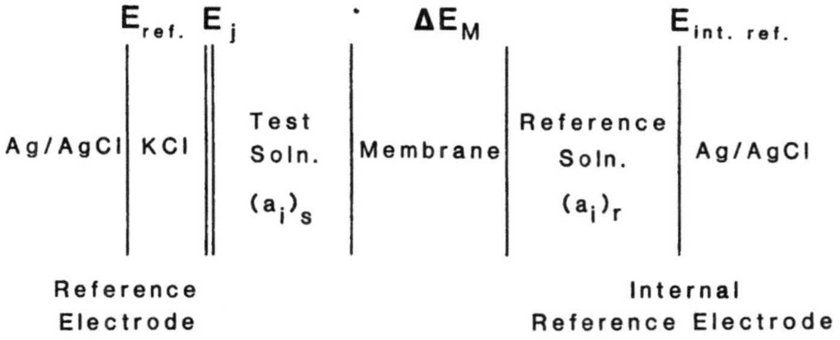

Fig. 6. Energetics associated with a complete cell used for ion measurements

3.2 Ion-Selective Membranes

Ion-selective membranes derive their permselective properties from either ion exchange, solubility or complexation phenomena. Current ion-selective electrodes contain membranes which consist of glass, solid or liquid phases.

Glass electrodes can be made which are selective for the hydrogen ion and alkali metal cations [9]. The silicate anions present in the hydrated layer of the glass act as fixed ion exchange sites. Selectivity is determined by ion exchange equilibria and the relative mobility of species in the hydrated glass. Hydrogen ion sensitive glass possesses a high selectivity for H^+ over Na^+ because of the high mobility of the hydrogen ion. Addition of Al_2O_3 to the glass increases the selectivity for alkali metal cations. Glass electrodes selective for hydrogen ions and sodium ions enjoy widespread commercial application.

Solid state membranes comprise either a single crystal or a precipitate of an insoluble salt. The single crystal membrane has interstitial vacancies with an appropriate size, shape and charge to permit permeation of a single ionic species. For example, lanthanum fluoride is extremely selective for the determination of fluoride ions [62]. Membranes comprised of an insoluble salt can be used to sense either the metal ion or the associated anion. The most common of these are the insoluble silver salts (AgCl, AgBr, AgI, Ag_2S) [62]. In these materials the silver ion is the mobile species and the sample/pellet interfacial potential develops due to a difference in the activity of Ag^+ in each phase. These electrodes can also be used to detect the anionic partner, since its presence in the sample will decrease the Ag^+ concentration. The selectivity coefficient for anions of equal charge is given by the quotient of their respective solubility products with Ag^+. The sensitivity is limited by the solubility of the membrane.

Liquid membranes consist of an organic phase, which by its hydrophobic nature is relatively impermeable to ions. Originally organic solvents such as decanol were used in conjunction with a porous hydrophobic membrane. These have been replaced by plasticized polyvinyl chloride membranes which behave like liquids yet have improved mechanical properties [50]. Other polymers such as silicone, polyurethane and urushi, a Japanese lacquer, have also been employed. The partitioning of ions into this phase is selectively aided by either ion exchangers or neutral carriers.

In general, an ion exchanger bears a charge opposite to the ion of interest and ion pair formation allows the transport of the ion through the membrane. The ion exchanger must be quite hydrophobic to be soluble in the organic phase. Quaternary ammonium cations have been used to prepare anion selective electrodes for chloride, nitrate, perchlorate and thiocyanate [62]. The order in which ions are preferred will be influenced by the ion pair formation constant as well as the partition coefficient of the ion in the hydrophobic phase. Therefore, membrane additives such as plasticizers may greatly influence the selectivity of such electrodes. Some ion exchangers are actually chelating agents and exhibit superior selectivity, the most familiar example being the organic phosphate type ion exchanger for the determination of calcium [63].

Neutral carriers are organic complexing agents which are capable of sequestering and transporting ionic species in a hydrophobic organic phase. The antibiotics, valinomycin and nonactin were the first neutral carriers to be incorporated in an ISE [64, 65]. These macrocyclic neutral carriers contain a polar internal cavity and an outer hydrophobic shell. The excellent selectivity exhibited by valinomycin for potassium ions is remarkable and is believed to be due to the perfect fit of potassium in the cavity. The search for other neutral carriers was then initiated and has been reviewed [61].

3.3 Electrode Fabrication

Traditionally, electrodes have either been constructed as dipping type for beaker studies or flow-through for continuous monitoring. The fabrication of glass electrodes is truly an art and glass blowing skills are required. An excellent compilation of step-by-step instructions for the construction of bulb type, flat ended and capillary glass electrodes is available [66]. However, many commercial electrodes are now available [67] which should satisfy most needs. Solid state electrodes are rather straightforward to make. The single crystal or pressed pellet must be sealed into some body so that there is no contact between the sample and reference solutions. The construction of PVC-based electrodes has recently been described [55]. A novel approach has been disclosed for impregnating silicone rubber tubing with the ion selective material [68]. This procedure is so simple, electrodes can be prepared virtually in ten minutes time.

The desire to miniaturize and simplify the construction of ion-selective electrodes has led to the development of solid contact devices. The liquid internal filling solution is replaced by a solid contact between the metal conductor and the ion-selective membrane. The internal filling solution contains a constant concentration of ions needed to maintain a constant potential at the reference wire and the inner surface of the membrane. In a solid contact arrangement, measure must be taken to ensure that a constant potential is preserved at the conductor/membrane interface. The construction of a suitable solid contact will differ for different types of ISE's and has been recently reviewed [69]. Initially, PVC membranes were coated onto a platinum wire [70], whereas solid contact glass electrodes were constructed with Ag paint or coated with molten AgCl [9]. Completely solid state sensors comprising pH glass, Na^+ glass or a LaF_3 crystal have been constructed with an AgF contact material [71]. In this case the need for a separate reference electrode with a junction is eliminated by operation in a differential mode. The pH-LaF_3 combination was demonstrated for fluoride analysis at a constant pH.

The application of CHEMFET's to the measurement of ions other than H^+ was briefly discussed in the last section. Such devices are commonly referred to as ion sensitive field effect transistors (ISFET) and have been the subject of much debate [72, 73]. The problem is that although pH-sensitive materials such as SiO_2, Si_3N_4 or Al_2O_3 may be easily fabricated as FET gate materials, more complex materials such as PVC/valinomycin for K^+ sensing are not easily integrated with traditional semiconductor processing techniques. In addition, the stability of the ion selective membrane/semiconductor interface is questionable and may result in considerable drift. These problems, long associated with ISFET's, may not be completely insurmountable. Commercial devices are just beginning to emerge for Na^+, K^+, Ca^+ and pH [19] and may be suitable for certain applications. However, it is not likely that ISFET's will be able to compete with conventional ISE's in the main clinical laboratory.

3.4 Clinical Application

In order to be useful in a clinical laboratory, an ion-selective electrode must meet very stringent requirements. The selectivity must be such that physiological levels

Table 1.

	Cations	Anions
Principal	Sodium	Chloride
	Potassium	Bicarbonate
Secondary	Calcium	Phosphate
	Magnesium	Sulfate
	Ammonium	Proteins
		Organic acids

of other ions do not contribute to the electrode response. The electrode response must be rapid to ensure a high throughput (60–120 samples/hour). The electrode must exhibit a long use life for cost and convenience considerations. Finally, the accuracy and precision requirements of the clinical laboratory must be met. The species in physiological fluids which bear a positive or negative charge are listed as principal and secondary electrolytes in Table 1. Currently ISE's are available for the determination of H^+, Na^+, K^+, Cl^-, Ca^{2+}, CO_3^{2-}, NH_4^+ and Li^+. The normal concentration range of these ions in serum is listed in Table 2 along with membrane type and selectivity coefficients.

Table 2.

Ion	Normal range	Membrane	Selectivity coefficient	Ref.
H^+	pH 7.33–7.45	Glass	Na^+ $< 10^{-13}$	9)
		TDA, PVC	Na^+ 4×10^{-11}	74)
Na^+	125–148 m*M*	Glass	H^+ 3.0 K^+ 10^{-3}	9)
		ETH 227, PVC	H^+ 0.1 K^+ 3×10^{-2}	75)
K^+	3.5–5.3 m*M*	Valinomycin, PVC	H^+ 10^{-4} Na^+ 10^{-4}	76)
Cl^-	98–106 m*M*	Ag/AgCl	Br^- 1.2 I^- 86.5	62)
		Aliquat 336, decanol	Br^- 3.0	62)
Ca^{2+}	2.12–2.60 m*M* (total)	ETH 1001, PVC	Na^+ 3.2×10^{-6} K^+ 4×10^{-6} Mg^{2+} 1.3×10^{-5}	77)
		$(RO_2)_2PO_2^-$ dioctylphenyl phosphonate, PVC ($R = C_8$–C_{16})	H^+ 10^{-1} Na^+ 1.6×10^{-3} K^+ 6×10^{-3} Mg^{2+} 1×10^{-2}	62)
Li	0.5–1.5 m*M* (therapeutic)	ETH 1810, PVC	Na^+ 3.5×10^{-3} K^+ 2.5×10^{-3} Ca^{2+} 2×10^{-3}	78)
NH_4^+	33–66 m*M*	Nonactin, PVC	Na^+ 2×10^{-3} K^+ 0.10	79)
CO_3^{2-}	21–28 m*M* (as HCO_3)	quaternary ammonium compound, p-butyl-trifluro-acetophenone, PVC		80)

The electrolytes Na^+, K^+ and Cl^- are second only to glucose in being the most frequently run hospital tests. Many clinical chemistry analyzers now contain an ISE "module" for electrolyte analysis. Most commonly the "module" will consist of a Na^+-glass electrode, a valinomycin/PVC electrode, a Ag/AgCl pellet or a quaternary ammonium ion/PVC electrode and a reference electrode. A selective electrode for the bicarbonate ion continues to elude workers in the field. An indirect measurement of HCO_3^- must be made. The sample is usually reacted with acid to evolve carbon dioxide gas which is measured with a traditional Severinghaus type CO_2 electrode. Alternatively, the sample is treated with base to convert HCO_3^- to CO_3^{2-} and a carbonate ion-selective electrode is used [80]. In this manner, the complete primary electrolyte profile is obtained electrochemically.

Calcium ion-selective electrodes have recently been commercialized for the measurement of either total or ionized calcium [77]. Approximately 45% of the calcium present in serum is bound to proteins, 5% is complexed to simple anions and 50% exists as the free ion. Traditionally, total calcium measurements have been made by releasing the protein bound fraction. An ion-selective electrode has now allowed the free (ionized) calcium to be measured directly. There has been much debate on the clinical significance of these measurements. The dependence of ionized calcium on pH must be considered. Samples must be either treated anaerobically, tonometered to a constant pH or have a correction factor applied.

It has been long believed that a lithium ion-selective electrode would render obsolete the flame photometer in the clinical laboratory. Lithium is administered to manic depressive psychiatric patients. Since the therapeutic range (0.5–1.5 m*M*) is quite close to the toxic range (>2 m*M*), it must be closely monitored. Most of the ionophores proposed to date have not met the Li^+/Na^+ selectivity required for an interference-free assay. However, it has been reported that calibration in the presence of 140 m*M* Na^+ permitted the analysis of Li^+ in serum [78]. The errors observed are due to fluctuations in the Na^+ concentrations in the sample. More selective ionophores would certainly improve the accuracy of this method.

Although serum ammonia levels are not routinely measured, it is a useful indicator of Reye's syndrome and should be monitored in newborns at risk of developing hyperammonemia [81]. Ammonia is produced in many analytically useful enzyme reactions and the ammonium ISE has been used as the base sensor in several enzyme electrodes (see next section). In addition to valinomycin, other antibiotics such as the nonactin homalogs and gramicidins also behave as ionophores. The nonactin homologs were originally studied for their ability to selectively bind potassium ions [79]. It was then discovered that ammonium ions were preferred over potassium ions, and the selectivity coefficient $K_{NH_4^+} = 0.12$ was reported. Since ammonia is present at fairly low levels in serum, this selectivity is not sufficient to to accurately measure NH_4^+ in the presence of K^+. An extra measure of selectivity can be gained by using a gas permeable membrane to separate the ammonia gas from the sample matrix [82].

3.5 Summary

Ion-selective electrodes have been well received in the clinical laboratory and have allowed electrolyte panels to be routinely run on automated analyzers. The convenience, low cost and speed of ISE technology has enabled it to supplant the conventional flame photometric method for the determination of sodium and potassium.

Ion-selective electrodes allow the measurement of ionic activity in diluted or undiluted whole blood, plasma or serum. The direct (undiluted) measurement may be preferred, since no sample pretreatment is necessary and the assay values are independent of hematocrit and amount of solids present. However, direct potentiometry by its very nature does not provide total concentration values similar to those obtained by flame photometry and indirect (diluted) potentiometry [83].

Future trends may include the commercialization of ISE's for other clinically significant ions such as bicarbonate, magnesium and phosphate. Solid contact electrodes and ISFET's may allow for mass production of smaller, less expensive devices. However, a high standard of performance must be achieved before conventional electrodes become obsolete.

4 Enzyme Electrodes

4.1 Introduction

The demand for monitoring common metabolites of diagnostic utility such as glucose, urea and creatinine continue to provide the impetus for a staggering research effort towards more perfect enzyme electrodes. The inherent specificity of an enzyme for a given substrate, coupled with the ability to electrochemically detect many of the products of enzymatic reactions initiated the search for "molecule-selective" electrodes.

The high specificity required for the analysis of physiological fluids often necessitates the incorporation of permselective membranes between the sample and the sensor. A typical configuration is presented in Fig. 7, where the membrane system comprises three distinct layers. The outer membrane, A, which encounters the sample solution is indicated by the dashed lines. It most commonly serves to eliminate high molecular weight interferences, such as other enzymes and proteins. The substrate, S, and other small molecules are allowed to enter the enzyme layer, B, which typically consist of a gelatinous material or a porous solid support. The immobilized enzyme catalyzes the conversion of substrate, S, to product, P. The substrate, product or a cofactor may be the species detected electrochemically. In many cases the electrochemical sensor may be prone to interferences and a permselective membrane, C, is required. The response time and sensitivity of the enzyme electrode will depend on the rate of permeation through layers A, B and C; the kinetics of enzymatic conversion; as well as the charac-

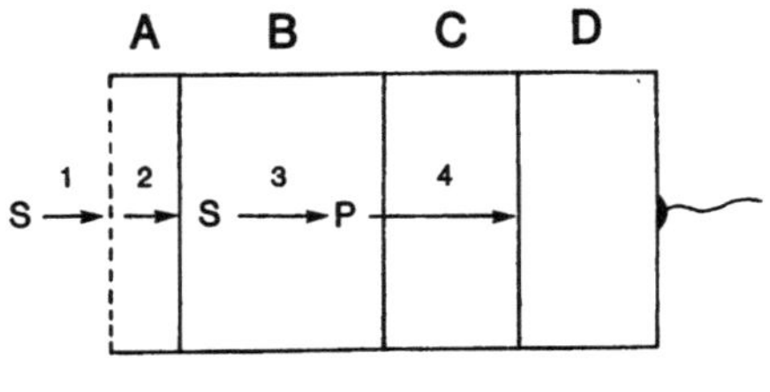

Fig. 7. Schematic diagram of an enzyme electrode. A, B and C comprise the membrane system, while D is the detector (either amperometric or potentiometric)

teristics of the electrochemical sensor itself. Theoretical models have been developed [84] and provide important insights in the design of enzyme electrodes.

4.2 Enzyme Immobilization

Enzyme immobilization methods have been extensively reviewed [84–86] and can be classified into three categories:
1) Carrier Binding
2) Cross-Linking
3) Entrapment

Carrier binding includes physical adsorption, ionic binding and covalent bonding of the enzyme to a solid support. Although more extensive synthetic reactions may be involved, covalent attachment provides the most stable linkage and is generally preferred. Popular supports include nylon [87–89], collagen [90], amine-modified porous glass [91], polyacrylamide [91,92], silica [92] and alumina [93].

The cross-linking method relies on bifunctional reagents to form intermolecular linkages between the enzyme molecules to render them insoluble. Often albumin is added as an extender and glutaraldehyde is most commonly employed. This material can then be either formed as a free standing membrane [94] or applied to the inner surface of the dialysis membrane [95].

The entrapment method is based on confining the enzyme within the lattice of a polymeric matrix. Polyacrylamide gels have successfully yielded stable enzyme films with a high retention of activity [96,97].

If the electrochemical sensor does not require a permselective membrane, immobilization of the enzyme onto the surface of the electrode is possible. Glassy carbon [98], graphite [99,100], reticulated vitreous carbon [97,101] and carbon paste electrodes [102] lend themselves to direct covalent attachment, adsorption and physical entrapment of the enzyme. Platinum electrodes may be aminosilylated to activate the surface for covalent bonding [103]. A novel physical entrapment approach relies on an electrode initiated polypyrrole film formation, accompanied by inclusion of the enzyme [104].

4.3 Applications

The enzymes which are compatible with electrochemical detection are the 1) oxidoreductases and 2) hydrolases. Oxidoreductases comprise all enzymes which catalyze oxidation-reduction reactions. Two major subsets exist within this category: oxidases and dehydrogenases. Flavin groups are responsible for the redox chemistry of the oxidases. The flavin prosthetic group is tightly bound by the apoenzyme and they function as an integral unit. The flavin oxidizes the substrate and an electron acceptor is required to return the flavin to its original oxidized state. A generalized reaction scheme is shown below. In biological systems, oxygen serves very efficiently as the

$$\text{Enz-FAD} + S_{red} \rightleftharpoons \text{Enz-FADH}_2 + S_{ox}$$
$$\text{Enz-FADH}_2 + O_2 \rightleftharpoons \text{Enz-FAD} + H_2O_2$$

electron acceptor and is reduced to hydrogen peroxide. The electrochemistry of free flavins is well documented [105], however less is known about the enzyme conjugate.

Table 3. Reduction potentials for some electron acceptors

	$E^{0\prime}$, V vs. SCE[a]
O_2/H_2O	0.816
$Fe(CN)_6^{3-}$	0.316
O_2/H_2O_2	0.295
1,4-Benzoquinone	0.293
2,6-Dichloroindophenol	0.217
Methylene Blue	0.011
$Ru(NH_3)_6^{3+}$	−0.214
FAD	−0.219
NAD^+	−0.320

[a] Midpoint potential at pH 7 vs. SCE taken from handbook of Biochemistry and Molecular Biology, p. 122, Boca Raton, Florida, CRC Press (1975)

Synthetic electron acceptors have been shown to react very rapidly with free flavins [104]. The combination of a flavin with an apoenzyme often inhibits the reaction with certain electron acceptors. The reaction with an electron acceptor will be thermodynamically favored if its standard reduction potential is larger than that of the flavin. A list of electron acceptors along with their reduction potentials can be found in Table 3.

Following the pioneering work of Clark and Lyons [107] as well as Updike and Hicks [108], electrochemical detection has been achieved in a variety of formats. The simplest relies on monitoring the consumption of oxygen with a traditional Clark-type oxygen sensor (Sect. 2.4) [96, 107–113]. In oxygen depletion, the difference between two large numbers is measured, while ideally, hydrogen peroxide detection would yield a positive signal on a zero background. All is not ideal, however, and hydrogen peroxide is oxidized at potentials where a number of other species normally present in serum are oxidized as well. Permselective cellulose acetate membranes have been developed for hydrogen peroxide which effectively screen out any anionic interferents such as ascorbic acid and uric acid. The permselectivity for H_2O_2 is also aided by the small size and rapid diffusion of H_2O_2 as compared to other species. The cellulose acetate must be quite thin to maintain rapid response time and can be difficult to manufacture and couple to the immobilized enzyme. Commercialization of a two-ply laminated membrane was achieved by Yellow Springs Instruments [114]. The membrane comprises 1) a support layer (6 μm polycarbonate, 0.03 μm pore size) which serves as a barrier to high molecular weight substances; 2) an adhesive layer of the enzyme (glutaraldehyde, enzyme-albumin matrix); and 3) a 1 μm homogeneous cellulose acetate layer. Other embodiments of this approach have been published [115–117].

Other solutions to dealing with interferences in the detection of H_2O_2 have included the use of a copper(II) diethyldithiocarbamate precolumn [9, 103] to oxidize the sample before it reaches the immobilized enzyme, as well as the use of a palladium/gold sputtered electrode which catalyzes the oxidation of hydrogen peroxide [118]. In addition, peroxidase has been used to catalyze the reaction between hydrogen peroxide and iodide [119, 120], ferrocyanide [121, 122] and organo-fluorine compounds [123]. Amperometric detection of ferricyanide is accomplished at lower potentials than H_2O_2

and is less prone to interferences. Iodide and fluoride are measured with ion-selective electrodes.

When oxygen is used as the electron acceptor, its low solubility ($\sim 125\ \mu M$ at RTP) often limits the linear range at high concentrations of substrate. In the case of glucose, accurate readings on diabetic patients are achieved by diluting the sample. However, it may be desirable to perform the analysis on whole blood (finger stick) or in-vivo (insulin pump). Towards this goal, a fine Pt screen and thin Au film electrodes have been developed to allow O_2 to freely diffuse into the enzyme layer from the back side [124]. In addition, a silanized dialysis membrane was employed to restrict the diffusion of glucose from the sample into the enzyme layer [117]. Catalase can be used in conjunction with the oxidase enzyme to extend the linear range [125]. The consumption of oxygen is reduced by 50%, according to the reactions detailed below.

$$\text{glucose} + O_2 \xrightarrow{\text{glucose oxidase}} \text{gluconic acid} + H_2O_2$$

$$H_2O_2 \xrightarrow{\text{catalase}} 1/2\ O_2 + H_2O$$

The use of alternate cofactors reduces the complexity of whole blood glucose measurements since the concentration of cofactor can be more easily controlled. Also, synthetic electron acceptors may be tailored such that a permselective membrane is not required for interference-free detection. Towards this goal, several chemically modified modified electrodes [126] containing both immobilized mediator and glucose oxidase have been evaluated. Ferricenium [127–130], N-methylphenazium [99], benzoquinone [102] and ruthenium compounds [131] have been successfully used as electron acceptors for glucose oxidase. The ferricenium electrodes require only a protective outer membrane, exhibit fair stability after pretreatment and provide linear response in the diabetic range. However, much work remains before acceptable clinical performance with physiological samples can be guaranteed [130].

Although the primary focus of oxidase based enzyme electrodes has been the determination of glucose, the list of extensions to other analytes is considerable. Systems have been described for cholesterol [123,132–136], galactose [114,135,137] uric acid [138,139], lactate [140–143], pyruvate [144], creatinine [145], serum lipase [146], ethanol [135,147] and amino acids [148].

Many dehydrogenase enzymes catalyze oxidation/reduction reactions with the aid of nicotinamide cofactors. The electrochemical oxidation of nicotinamide adenine dinucleotide, NADH, has been studied in depth [149–154]. The direct oxidation of NADH has been used to determine concentration of ethanol [155–157,162], lactate [155–157,160,162,163], pyruvate [157], glucose-6-phosphate [161], lactate dehydrogenase [158,159,161] and alanine [164]. The direct oxidation often entails such complications as electrode surface pretreatment, interferences due to electrode operation at very positive potentials, and electrode fouling due to adsorption. Subsequent reaction of the NADH with peroxidase [165–168] allows quantitation via the well established Clark electrode. Ferricyanide has been employed as an electron acceptor for lactate in place of NAD^+ [169]. Alternatively, diaphorase can be used in conjunction with ferricyanide [170], Bindschedler's Green [171], ferricenium [136] or dichlorophenylindophenol [172] to catalyze the oxidation of NADH while providing a more easily monitored species than NADH itself.

The oxidation of NADH has been mediated with chemically modified electrodes [126] whose surface contains synthetic electron transfer mediators. The reduced form of the mediator is detected as it is recycled electrochemically. Systems based on quinones [173-175], dopamine [176], chloranil [177], 3-β-napthoyl-Nile Blue [178], phenazine methosulphate [179], meldola blue [180, 181] and similar phenoxazines [182, 183] have been described. Conducting salt electrodes consisting of the radical salt of 7,7,8,8-tetracyanoquinodimethane and the N-methylphenazium ion have been reported to show catalytic effects [184-186]. The main drawback to this approach is the limited stability of the modified electrodes. The meldola blue electrode could be easily renewed on a daily basis and was successfully used in the determination of glucose with immobilized glucose dehydrogenase [181].

A recently characterized class of dehydrogenases are the quinoproteins which contain a pyrroloquinolene quinone prosthetic group and do not require a separate cofactor [187]. Electron transfer mediators such as phenazine ethosulphate [188], 2,6-dichloroindophenol [188] and ferricenium ions [189] have been used to recycle the quinoprotein; the reduced mediator is detected amperometrically.

Hydrolase enzymes catalyze the hydrolysis of a substrate and are most commonly coupled with potentiometeric electrodes [55]. The pioneering work in this field focussed on developing an enzyme electrode for the determination of urea. Urease catalyzes the hydrolysis of urea to ammonium and bicarbonate ions according to the reaction detailed below.

$$H_2N-\overset{\overset{\displaystyle O}{\|}}{C}-NH_2 + 2\,H_2O + H^+ \xrightarrow{\text{urease}} 2\,NH_4^+ + HCO_3^-$$

Electrochemical detection has been achieved in a number of ways. The change in pH has been sensed with a traditional glass pH electrode [190-192], an antimony electrode [193, 194] or amperometrically via the pH sensitive oxidation of hydrazine [195]. The ammonium ion concentration may be monitored with a cation selective glass electrode [196] or an neutral carrier based ion-selective electrode [197-199]. The pH may be adjusted so that either ammonia [113, 200-203] or carbon dioxide [204] may be detected using potentiometric gas sensors for these species. An ammonia sensitive semiconductor has been also applied to the determination of urea [205].

The development of an enzyme electrode for creatinine has been considerably more difficult. Creatinine iminohydrolase has been employed to selectively produce ammonia from creatinine [206-210]. The normal physiological concentration of creatinine is 50 μM, which may be equal or be even less than the levels of endogenous ammonia found in serum. The endogenous ammonia has been measured prior to enzymatic conversion [208] as well as enzymatically removed [209, 210].

Enzyme electrodes for the amino acids tyrosine [204, 211] and lysine [212] have been constructed by immobilizing their respective decarboxylases on a CO_2 gas sensing electrode. Adenosine deaminase, together with an NH_3 gas sensing electrode formed the basis for an enzyme electrode for the ribonucleoside adenosine [213].

4.4 Summary

As is evident from the numerous publications on enzyme electrodes in the literature, a tremendous effort has been expended in this field resulting in a long list of applications for diagnostically significant analytes. To date widespread commercialization of enzyme electrodes has not been realized. The manufacturing of the multilayer composite membranes often required is not trivial and performance requirements such as long use life and shelf life are not easily met. Recent developments in chemically modified electrodes may simplify the oxidoreductase-based systems from a manufacturing standpoint and such devices may be applicable in a disposable format. Commercialization of the hydrolase-based enzyme electrodes may be awaiting refinements in the response characteristics of potentiometric gas sensors, which at the present time severely limit their throughput. New developments in the area of solid contact ISE's and ISFET's may reduce the cost to manufacture the underlying potentiometric sensor, so that hydrolase-based enzyme electrodes could also see commercial application in a disposable format.

5 Electrochemical Immunoassay

5.1 Introduction

One of the most exciting new applications for electrochemistry in the last decade has been in the area of immunoassay. With more than a hundred million immunoassays being performed world-wide each year, researchers have begun to carve out new immunoassay strategies which exploit the excellent detection limits that can be achieved with modern electrochemical techniques.

Broadly speaking, electrochemical immunoassay (ECIA) methods can be divided into two categories, direct and indirect. In the former case, a potentiometric electrode is used as a direct probe of the antigen-antibody binding event, Fig. 8. All other methods are indirect, since they rely on labels bound to either antigen or antibody molecules in order to generate an electrochemical signal. This electrochemical signal may be potentiometric or amperometric. Indirect ECIA based on amperometric detection is receiving much attention and is the subject of a recent excellent review [214]. All methods under active study are discussed here.

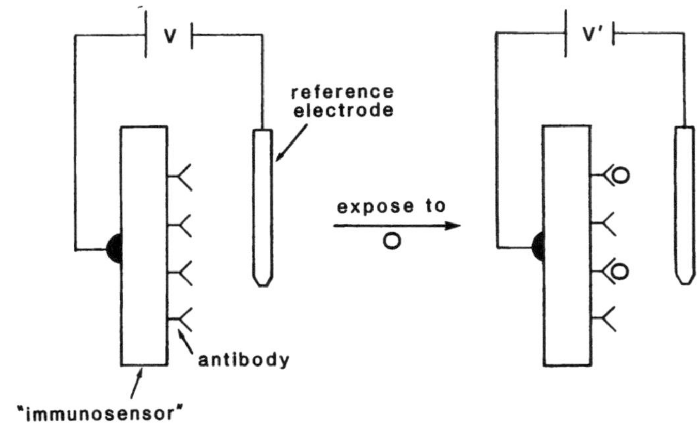

Fig. 8. Representation of a direct potentiometric immunoassay

5.2 Direct ECIA

The simplest ECIA format is one which employs no labels or separation steps to detect the primary immunochemical reaction of interest. The groundwork was laid for this "direct" approach to ECIA in 1966 [215] and 1968 [216] with the demonstration of changes in transverse electrical impedance of antibody-containing lipid films induced by antigen binding. In 1975 the term "immunoelectrode" was coined in the literature [217], and although the electrode reported was not based on an immunochemical reaction, the potentiometric response observed for the concanavalin A/polysaccharide model system provided much impetus for further investigations [218–227] in the area. Despite the nearly two decades of research, however, no direct ECIA systems have achieved commercial success. In addition, two recent manuscripts have been so bold as to imply that an equilibrium or even steady-state direct reading potentiometric immunosensor useful in complex media is impossible to design [225, 227].

While a detailed discussion of the potential-generating mechanisms in these systems is necessary to fully understand the difficulty of implementing practical systems, a brief explanation is that while adsorption of protein at an electrode surface can in of itself cause a change in interfacial potential due to changes in surface charge, the changes in potential are mainly due to alterations in the ion-exchange processes for small inorganic ions at the interface which are caused by the protein adsorption event. The desirable use of complex media such as serum or urine is not possible with such a system since the variation in concentration of numerous small ions in clinical samples will cause sample-to-sample variation of the response of the electrode to the species of interest. Even more importantly, the technique has not demonstrated the ability to reliably detect analytes below $\sim 10^{-7}$ M and so would not be useful for a large number of important tests.

In response to the apparent lack of success of the simple approaches to direct ECIA, there has emerged a more sophisticated technique [228–231] which strictly speaking should be classified as "indirect". This method uses the antibody-antigen binding event to modulate a background potential generated by a fixed marker ion, Fig. 9. Specifically, an antigen corresponding to an antibody to be measured is coupled to an ionophore to form an antigen-carrier conjugate. The conjugate is incorporated into a support membrane which is mounted in the sensing tip of a conventional potentio-

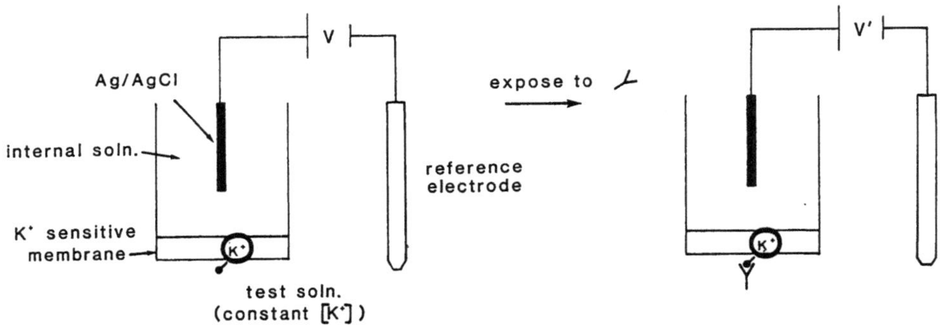

Fig. 9. Representation of the use of the modification of an ionophore's potentiometric response in order to detect antibody binding. A constant ion activity (in this case K^+) must be maintained in the sample solution

metric membrane electrode. The resulting electrode is exposed to a constant activity of marker ion compatable with the ionophore portion of the conjugate to produce a stable background potential. When antibody is added to the background electrolyte a potential change proportional to the antibody concentration is produced. The sensitivity of the technique is in the nanomolar range, and thus could be useful in some applications. One drawback to the approach, however, is that a constant marker ion concentration must be maintaned in all samples. In addition, a more complex competitive binding format is necessary in order to measure antigens since upon binding they are likely to have little effect on ionophores coupled to antibodies because antibodies are resonably large (MW > 100,000) molecules. Nevertheless, the technique represents an important addition to the arsenal of ECIA approaches.

5.3 Indirect ECIA

Due to the apparent inadequacies of direct methods, much effort has been devoted to "indirect" immunoassay methods (requiring labels and/or separation steps) which draw heavily from advances [232] made in enzyme immunoassay (EIA) in recent years. EIA's requiring wash steps are generally referred to as heterogeneous immunoassays, the most important of which is the enzyme-linked immunosorbent assay (ELISA). The most important homogeneous (no wash steps) EIA is the enzyme-multiplied immunoassay technique (EMIT[a]). In fact, many indirect ECIA methods are simply EMIT of ELISA assays which rely on enzyme generated products that can be detected electrochemically instead of optically, Fig. 10. Other methods employ nonenzymatic electroactive labels which are detected either directly at an electrode or indirectly by coupling with enzymes to generate other electrode-detectable species. Potential advantages of the electrochemical approach are less complex instrumentation and increased dynamic range [233], vide infra.

Perhaps the earliest reports to appear in the area of ECIA were those of Breyer and Radcliff [234, 235]. In their homogeneous assay an aromatic diazonium ion was coupled to egg albumin antigen to form an electroactive label whose amperometric response was shown to be modulated by antibody binding. Some 20 years later, reports [236, 237] appeared which described the use of a silver sulfide ion-selective electrode in somewhat cumbersome formats to monitor antibody-antigen reactions by measuring the degree of protein denaturation. Later reports described the use of iodide [238] and ammonia [239] potentiometric electrodes to perform "sandwich" type enzyme-coupled heterogeneous assays. A fluoride ion-selective electrode has been used [240] for IgG determination in human serum by employing horseradish peroxidase as an antibody label. Fluoride ion is generated by HRP-catalyzed oxidation of *p*-fluoroaniline by hydrogen peroxide. In an early amperometric method [241] catalase was used as an enzyme label in a competitive heterogeneous assay format to generate oxygen from hydrogen peroxide substrate. The oxygen was then detected at a Clark-type electrode. This same approach has been used to demonstrate alpha-fetoprotein [242], theophylline [243] and human serum albumin assays [244]. A simpler version of this where the labeled antigen (in this case an antigen analog of lower affinity than the sample antigen) is already bound

[a] EMIT is a U.S. registered trademark of Syva Co., Palo Alto, California.

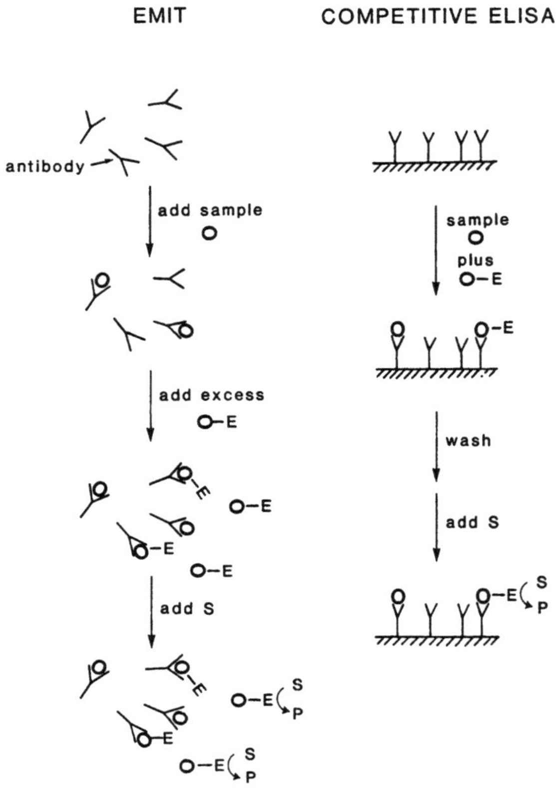

Fig. 10. Comparison of a typical homogeneous immunoassay (left) with a heterogeneous format (right). Depending on choice of reagents, there are many possible variations on these basic schemes

to the antibody-activated membrane before exposure to the sample antigen has demonstrated $\sim 10^{-9}$ *M* sensitivity [245].

Very recently, a "sandwich" assay [233] for prostatic acid phosphatase antigen was carried out using two cascaded enzyme reactions to provide amplification of the immunochemical event. In one format, an optical readout was used whereby a formazan dye was generated by reaction of a dye precursor and NADH generated from the second enzyme cycle. In the electrochemical format, the NADH generated in the second enzyme cycle was used to reduce $Fe(CN)_6^{3-}$ to $Fe(CN)_6^{4-}$ which was then detected amperometrically. While the use of $Fe(CN)_6^{3-/4-}$ in ECIA has appeared in the literature previously [246], the results reported for this "sandwich" assay are important because they show that for this case the electrochemical format has superior dynamic range over the optical method and requires simpler instrumentation. In addition, excellent sensitivity (near 10^{-12} *M*) was demonstrated for both methods. The slightly lower precision observed in the electrochemical format was ascribed to non-optimal design of the experimental apparatus. Amperometric detection with a "sandwich" assay has also been used to demonstrate moderate sensitivity ($\sim 10^{-8}$ *M*) for detection of HCG in serum by use of a ferrocene derivative as an electron transfer mediator to measure enzyme activity [247].

While adequate sensitivity can be achieved with some of the indirect techniques described above, they generally require a separation step during the analytical procedure. This is undesirable, and so there have also appeared reports which attempt to exploit homogeneous immunoassay formats with electrochemical detection. Like other homogeneous assays, homogeneous ECIA systems usually have poorer sensitivity than heterogeneous assays, and are more likely to suffer from interferences from other sample constituents normally removed in the separation step.

The simplest homogeneous ECIA formats have employed an amperometrically detectable antigen label. The pioneering work for this approach was again provided by Breyer and Radcliff [234, 235], vide supra. Since this early work, ferrocene has been used to label morphine for detection at a glassy carbon electrode in a flow system [248]. Amplified detection of a ferrocene label by using it as a cofactor in the oxidation of glucose by glucose oxidase has been used to detect lidocaine antigen at the micromolar level [249]. Detection of esteriol has been accomplished [250] by using nitro groups to impart electroactivity to this molecule. Differential pulse polarography was used to monitor the free labeled estriol. Mercuric acetate has also been used to label estriol [251]. The problem with the approach is that complex media such as blood and urine contain other electroactive species which interfere and limit the practical sensitivity.

Complement induced immune lysis of cells and liposomes to release markers which are then detected electrochemically has been used to detect antibodies and antigens in a homogeneous format at nanomolar levels [252–256]. Both amperometric and potentiometric electrodes have been employed. Unfortunately, major improvements in sensitivity appear unlikely, and instability of liposomes makes development of stable reagents for commercial systems difficult.

A fundamentally new homogeneous enzyme immunoassay system has recently been described which makes use of an electrochemical read-out [257]. The assay uses an antigen-coupled ammonia liberating enzyme along with two antibodies. Deaminating enzymes had been used previously to demonstrate ECIA [258], but in this case an anti-enzyme antibody inhibits the enzyme while the antigen-selective antibody reverses the inhibition process. When samples containing free antigen are present in the assay mixture, there is competition for the anti-antigen antibody sites and protection against anti-enzyme antibodies is diminished. The extent of enzymatic reaction is monitored with an ammonium ion-selective electrode. The preliminary results are promising, but improvements in detection limits and assay times are necessary before this approach can become viable.

Homogeneous approaches based on the more conventional EMIT technology have also been investigated as candidates for electrochemical read-out schemes. Production of NADH in a conventional EMIT format for antigen determination using glucose-6-phosphate dehydrogenase (G6PD) has been monitored electrochemically in a flow system and shows good correlation with the normal optical procedure used for NADH determination [259]. Electrode fouling and interferences from other electroactive species are problems which need to be addressed in this approach. The same G6PD/NADH system coupled with amperometric NADH detection has also been used to demonstrate subnanomolar detection of antibodies [260].

A fairly sensitive ($\sim 10^{-8}$ *M*) homogeneous ECIA technique for human IgG using chloroperoxidase catalyzed CO_2 production and subsequent potentiometric detection has recently been reported [261]. A more complex scheme using enzymes and amperometric determination of H_2O_2 has demonstrated micromolar sensitivity [262].

5.4 Miscellaneous Techniques

There are some emerging techniques for immunoassay which rely in part on electrochemistry but are not conveniently categorized with the techniques described above. The most important of these new techniques is the so-called electrochemiluminescence immunoassay (ECLIA). Fundamentally, this approach is a hybrid of electrochemical and optical methods. First described in 1981 [263], the technique has its roots in chemiluminescence immunoassay [264]. In chemiluminescence immunoassay a label capable of emitting light upon suitable chemical activation is employed. The ECLIA approach uses an electrode to activate light emission from a label, thus potentially lowering the number of chemical reagents which must be used and simplifying the entire assay. A rather ill-defined and non-optimized homogeneous assay system has recently demonstrated micromolar sensitivity using an aminopyrene label [265]. In addition, Ru(2,2'-bipyridine)$_3^{3+}$ has recently been studied for its potential as a label in such systems, and detection of this complex at concentrations as low as 10^{-13} *M* was demonstrated [266]. These results are exciting, and will likely lead to much more intensive investigation of the approach in the near future.

5.5 The Future of ECIA

Based on many of the advances described above in electrochemical approaches to immunoassay, it is tempting to conclude that commercialization of some of these approaches is imminent. This may be true, but the historical use of optical methods for many clinical chemistry tests coupled with their rapidly growing use in immunoassay is a difficult barrier for any radically different method to overcome, though electrochemical sensors have become more important in the clinical chemistry laboratory over the last decade. In any event, to be successful ECIA methods will have to demonstrate clear superiority over existing and emerging technologies in both cost and performance. Some of the more recently described approaches such as those using enzyme amplified amperometric detection [233, 249] and ECLIA [263, 265, 266] appear most promising because they offer some clear advantages over similarly formatted optical methods.

6 References

1. Davis, J. E., Solsky, R. L., Giering, L., Malhotra, S.: Anal. Chem. *55*, 202R (1983)
2. Czaban, J. D.: ibid. *57*, 345A (1985)
3. Rechnitz, G. A.: Trends in Anal. Chem. *5*, 172 (1986)
4. Koryta, J.: Electrochim. Acta *31*, 515 (1986)
5. Pruden, E. L., Siggaard-Anderson, O., Tietz, N. W.: Blood Gases and pH, in: Textbook of Clinical Chemistry (ed.) Tietz, N. W. p. 1191, Philadelphia, Saunders 1986
6. For a reasonably recent discussion of some specific modern blood gas instrumentation, see: Winckers, E. K. A., Teunissen, A. J., Van Den Camp, R. A. M., Maas, A. H. J., Veefkind, A. H.: J. Clin. Chem. Clin. Biochem. *16*, 175 (1978)
7. For a detailed discussion of the art of sample handling and data acquisition for blood gas measurements, see: National Committee for Clinical Laboratory Standards. Blood gas pre-analytical considerations: specimen collection, calibration, and controls; Proposed Guideline. NCCLS publication C27-P, Villanova, Pennsylvania, NCCLS, 1985

8. Bates, R. G.: Determination of pH: Theory and Practice, New York, Wiley 1964
9. Eisenman, G., ed.: Glass Electrodes for Hydrogen and Other Cations-Principles and Practice, New York, Dekker 1967
10. Westcott, C. C.: pH Measurements, New York, Academic Press 1978
11. Cobbold, R. S.: Transducers for Biomedical Measurements, p. 349, New York, Wiley 1974
12. Geddes, L. A., Baker, L. E.: Principles of Applied Biomedical Instrumentation, p. 125, New York, Wiley 1968
13. Ref. 10, pgs. 361–371
14. Bergveld, P.: IEEE Trans. Biomed. Eng. *BME-17*, 70 (1970)
15. Cheung, P. W., Fleming, D. G., Ko, W. H., Neuman, M. R., eds.: Theory, Design and Biomedical Applications of Solid State Chemical Sensors, West Palm Beach, Florida, CRC Press 1978
16. Sze, S. M.: Physics of Semiconductor Devices, New York, Wiley, 1981[2]
17. Caras, S., Janata, J.: Anal. Chem. *52*, 1935 (1980)
18. Van der Spiegel, J., Lauks, I., Chen, P., Babic, D.: Sensors and Actuators *4*, 291 (1983)
19. Sibbald, A., Covington, A. K., Carter, R. F.: Clin. Chem. *30*, 135 (1984)
20. Sibbald, A.: Sensors and Actuators *7*, 23 (1985)
21. Bergveld, P.: Biosensors *2*, 15 (1986)
22. Pace, S. J., Jensen, M. A.: Proceedings of the 2nd International Meeting on Chemical Sensors, p. 557, Bordeaux, France, July 7–10, 1986
23. Stow, R. W., Baer, R. F., Randall, B. F.: Arch. Phys. Med. *38*, 646 (1957)
24. Severinghaus, J. W., Bradley, A. F.: J. Appl. Physiol. *13*, 515 (1958)
25. Ref. 11, p. 399
26. Van der Schoot, B., Bergveld, P.: Sensors and Actuators *8*, 11 (1985)
27. Giner, J.: Electrochim. Acta *8*, 857 (1963)
28. Smith, P. T., Jordon, J.: Proc. Chem. Soc. *1960*, 246 (1960)
29. Roberts, J. L., Sawyer, D. T.: J. Electroanal. Chem. *9*, 1 (1965)
30. Severinghaus, J. W., Weiskopf, R. B., Nishimura, M., Bradley, A. F.: J. Appl. Physiol. *31*, 640, (1971)
31. Albery, W. J., Barron, P.: J. Electroanal. Chem. *138*, 79 (1982)
32. Albery, W. J.: U. S. Patent 4,377,446 (1983)
33. Price, J. F.: Extended Abstracts of the 169th Meeting of The Electrochemical Society, Boston, Massachusetts, Volume 86-1, p. 168 (1986)
34. Fatt, I.: Polarographic Oxygen Sensors, Cleveland, Ohio, CRC Press 1976
35. Hitchman, M. L.: Measurement of Dissolved Oxygen, New York, Wiley 1978
36. Giquere, P. A., Lauzier, L.: Can. J. Res. Sect. B *23*, 76 (1945)
37. Clark, L. C., Trans. Am. Soc. Artif. Intern. Organs *2*, 41 (1956)
38. Kreuzer, F., Kimmich, H. P., Brezina, M.: Polarographic Determination of Oxygen in Biological Materials in: Medical and Biological Applications of Electrochemical Devices (ed.) Koryta, J. Chapt. 6, New York, Wiley 1980
39. See Ref. 38, p. 178 for an excellent discussion of catheter sensors
40. Siu, W., Cobbold, R. W. C.: Med. Biol. Eng. *14*, 109 (1976)
41. Kimmich, H. P., Kuypers, M. H., Engles, J. M. L., Maas, H. G. R.: Adv. Exp. Med. Biol. *159*, 83 (1983)
42. Koudelka, M.: Sensors and Actuators *9*, 249 (1986)
43. Pace, S. J., Sarzycki, P. O., McKeever, R. T., Pelosi, L.: Digest of Technical Papers, 3rd International Conference on Solid-State Sensors and Actuators, p. 406, Philadelphia, Pennsylvania, June 11–14, 1985
44. Karagounis, V., Lun, L., Liu, C.-C.: IEEE Trans. Biomed. Eng. *BME-33*, 108 (1986)
45. Ref. 38, p. 186
46. Koryta, J.: Anal. Chim. Acta *183*, 1 (1986)
47. Arnold, M. A., Solsky, R. L.: Anal. Chem. *58*, 84R (1986)
48. Buck, R. P. in: Comprehensive Treatise on Electrochemistry, vol. 8, (ed.) White, R. E. p. 137 New York, Plenum Press 1984
49. Oggenfuss, P., Morf, W. E., Oesch, U., Amman, D., Pretsch, E., Simon, W.: Anal. Chim. Acta *180*, 289 (1986)
50. Thomas, J. D. R.: ibid. *180*, 189 (1986)

51. Moody, G. J., Thomas, J. D. R.: Ion-Selective Electrode Rev. *6*, 209 (1984)
52. Arnold, M. A., Meyerhoff, M. E.: Anal. Chem. *56*, 20R (1984)
53. Karube, I., Suzuki, S.: Ion-Selective Electrode Rev. *6*, 15 (1984)
54. Havas, J.: Ion- and Molecule-Selective Electrodes in Biological Systems, p. 91–167, Berlin Heidelberg New York, Springer-Verlag 1985
55. Meyerhoff, M. E., Opdycke, W. N.: Ion-Selective Electrodes, in: Advances in Clinical Chemistry *25*, 1–47 New York, Academic Press 1986
56. Savory, J., Bertholf, R. L., Boyd, J. C., Bruns, D. E., Felder, R. A., Lovell, M., Shipe, J. R., Wills, M. R., Czaban, J. D., Coffey, K., O'Connell, K. M.: Anal. Chim. Acta *180*, 99 (1986)
57. Kessler, M., Hofer, J., Harrison, D. K. (eds.): Ion Measurement in Physiology and Medicine, Berlin Heidelberg New York, Springer-Verlag 1985
58. Simon, W., Ammann, D., Anker, U., Oesch, U., Band, D. M.: Ann. N.Y. Acad. Sci. *428*, 279 (1984)
59. Purdy, W. C.: Chem. Int. *4*, 13 (1984)
60. Ladenson, J. H.: Anal. Proc. *20*, 554 (1983)
61. Oggenfuss, P., Morf, W. E., Oesch, U., Ammann, D., Pretsch, E.: Anal. Chim. Acta *180*, 299 (1986)
62. Koryta, J.: Ion-Selective Electrodes, Cambridge, Cambridge University Press 1975
63. Moody, G. J., Thomas, J. D. R.: Ion-Selective Electrode Rev. *1*, 3 (1979)
64. Stefanac, Z., Simon, W.: Chimia *20*, 436 (1966)
65. Stefanac, Z., Simon, W.: Microchem. J. *12*, 125 (1967)
66. Ref. 9, page 241
67. Anal. Chem. Lab Guide *58*, 88 (1986)
68. Fogt, E. J.: U.S. Patent 4,565,665 (1986)
69. Nikolskii, B. P., Materova, E. A.: Ion-Selective Electrode Rev. *7*, 3 (1985)
70. Cattrall, R. W., Freiser, H.: Anal. Chem. *43*, 1905 (1971)
71. Fjeldly, T. A., Nagy, K.: Sensors and Actuators *8*, 261 (1985)
72. Janata, J.: ibid. *4*, 255 (1983)
73. Meyerhoff, M.: Anal. Chem. *57*, 1995 (1985)
74. Schulthess, P., Shijo, Y., Pham, H. V., Pretsch, E., Ammann, D., Simon, W.: Anal. Chim. Acta *131*, 111 (1981)
75. Steiner, R. A., Oehme, M., Ammann, D., Simon, W.: Anal. Chem. *51*, 351 (1979)
76. Oehme, M., Simon, W.: Anal. Chim. Acta *86*, 21 (1976)
77. Oehme, M., Kessler, M., Simon, W.: Chimia *30*, 204 (1976)
78. Metzger, E., Ammann, D., Asper, R., Simon, W.: Anal. Chem. *58*, 132 (1985)
79. Scholer, R. P., Simon, W.: Chimia *24*, 372 (1970)
80. Scott, W. J., Chapoteau, E., Kumar, A.: Clin. Chem. *32*, 137 (1986)
81. Cooke, R. J., Jensen, R. L.: ibid. *29*, 867 (1983)
82. Meyerhoff, M. E.: Anal. Chem. *52*, 1532 (1980)
83. Arnold, M. A., Meyerhoff, M. E.: ibid. *56*, 20R (1984)
84. Carr, P. W., Bowers, L. D.: Immobilized Enzymes in Analytical and Clinical Chemistry, New York, Wiley Interscience Press 1980
85. Chibata, I.: Immobilized Enzymes, Research and Development, New York, John Wiley & Sons 1978
86. Guilbault, G.: Analytical Uses of Immobilized Enzymes, New York, Marcel Dekker 1984
87. Horby, W. E., Morris, D. L.: Immobilized Enzymes, Antigens, Antibodies and Peptides (ed.) Weetall, H. H., New York, Marcel Dekker 1975
88. Mascini, M., Iannello, M., Palleschi, G.: Anal. Chim. Acta *146*, 135 (1983)
89. Campanella, L., Tomassetti, M., Rapuoli, B., Bruni, M. R.: Am. Clin. Prod. Rev. June, 42 (1986)
90. Bardeletti, G., Coulet, P. R.: Anal. Chem. *56*, 591 (1984)
91. Masoom, M., Townshend, A.: Anal. Chim. Acta *166*, 111 (1984)
92. Yao, T., Sato, M., Kobayashi, Y., Wasa, T.: ibid. *165*, 291 (1984)
93. Watson, B., Stifel, D. N., Semersky, F. E.: ibid. *106*, 233 (1979)
94. Lubrano, G. L., Guilbault, G. G.: ibid. *97*, 229 (1978)
95. Mullen, W. H., Churchouse, S. J., Vadgama, P. M.: Analyst *110*, 925 (1985)
96. Hicks, G. P., Updike, S. K.: Anal. Chem. *38*, 726 (1966)
97. Lange, M. A., Chambers, J. Q.: Anal. Chim. Acta *175*, 89 (1985)

98. Bourdillon, C., Bourgeois, J. P., Thomas, D.: J. Am. Chem. Soc. *102*, 4231 (1980)
99. Jonsson, G., Gorton, L.: Biosensors *1*, 355 (1985)
100. Ianniello, R. M., Lindsay, T. J., Yacynych, A. M. 4: Anal. Chim. Acta *141*, 23 (1982)
101. Wieck, H. J., Heider, G. H., Yacynych, A. M.: ibid. *158*, 137 (1984)
102. Ikeda, T., Katasho, I., Senda, M.: Anal. Sci. *1*, 455 (1985)
103. Yao, T.: Anal. Chim. Acta *153*, 175 (1983)
104. Foulds, N. C., Lowe, C. R.: J. Chem. Soc., Faraday Trans. 1, *82*, 1259 (1986)
105. Gorton, L., Johansson, G.: J. Electroanal. Chem. *113*, 151 (1980)
106. Dixon, M.: Biochim. Biophys. Acta *226*, 269 (1971)
107. Clark, L. C., Lyons, L.: Ann. New York Acad. Sci. *102*, 29 (1962)
108. Updike, S., Shults, M. C., Busby, M.: J. Lab. Clin. Med. *93*, 518 (1979)
109. Havas, J., Porjesz, E., Nagy, G., Pungor, E.: Hung. Sci. Inst. *49*, 53 (1980)
110. Romette, J.-L., Froment, B., Thomas, D.: Clin. Chim. Acta *95*, 249 (1979)
111. Sokol, L., Garber, C., Shults, M., Updike, S.: Clin. Chem. *26*, 89 (1980)
112. Gough, D. A., Lucisano, J. Y., Tse, P. H. S.: Anal. Chem. *57*, 2351 (1985)
113. Mascini, M., Palleschi, G.: Anal. Chim. Acta *145*, 213 (1983)
114. Taylor, P. J., Kmetec, E., Johnson, J. M.: Anal. Chem. *49*, 789 (1977)
115. Tsuchida, T., Yoda, K.: Enzyme Microb. Technol. *3*, 326 (1981)
116. Tsuchida, T., Yoda, K.: Clin. Chem. *29*, 51 (1983)
117. Mullen, W. H., Keedy, F. H., Churchouse, S. J., Vadgama, P. M.: Anal. Chim. Acta *183*, 59 (1986)
118. Gorton, L.: ibid. *178*, 247 (1985)
119. Nagy, G., Von Storp, L. H., Guilbault, G. G.: ibid. *66*, 443 (1973)
120. Llenado, R. A., Rechnitz, G. A.: Anal. Chem. *45*, 826, 2165 (1973)
121. Kulys, J. J., Pesliakiene, M. V., Samalius, A.: Bioelectrochem. Bioenerg. *8*, 81 (1981)
122. Yao, T., Sato, M., Kobayaski, Y., Wasa, T.: Anal. Chim. Acta *165*, 291 (1984)
123. Siddiqi, I. W.: Clin. Chem. *28*, 1962 (1982)
124. Lobel, E., Rishpon, J.: Anal. Chem. *53*, 51 (1981)
125. Kondo, T., Ito, K., Ohkura, K., Ito, K., Ikeda, S.: Diabetes Care *5*, 218 (1982)
126. Murray, R. W.: Chemically Modified Electrodes. Electroanalytical Chemistry *13* (ed.) Bard, A. J., p. 411, New York, Marcel Dekker 1984
127. Cass, A. E. G., Davis, G., Grancis, G. D., Hill, H. A. O., Aston, W. J., Higgins, I. J., Plotkin, E. V., Scott, L. D. L., Turner, A. P. F.: Anal. Chem. *56*, 667 (1984)
128. Turner, A. P. F., Pickup, J. C.: Biosensors *1*, 85 (1985)
129. Lange, M. A., Chambers, J. Q.: Anal. Chim. Acta *175*, 89 (1985)
130. Claremont, D. J., Penton, C., Pickup, J. C.: J. Biomed. Eng. *8*, 327 (1986)
131. Crumbliss, A. L., Hill, H. A. O., Page, D. J.: J. Electroanal. Chem. *206*, 327 (1986)
132. Satoh, I., Karube, I., Suzuki, S.: Biotech. Bioeng. *19*, 1905 (1977)
133. Bertrand, C., Conlet, P. R., Gautheron, D. C.: Anal. Chim. Acta *126*, 307 (1981)
134. Bertrand, C., Conlet, P. R., Gautheron, D.: Anal. Lett. *12*, 1477 (1979)
135. Wollenberger, U., Kuhn, M., Scheller, F., Deppmeyer, V., Janchen, M.: Bioelectrochem. Bioenerg. *11*, 307 (1983)
136. Green, M. J., Hill, H. A. O.: J. Chem. Soc., Faraday Trans. 1, *81*, 1237 (1986)
137. Johnson, J. M., Halsall, H. B., Heineman, W. R.: Anal. Chem. *54*, 1394 (1982)
138. Nanjo, M., Guilbault, G. G.: Anal. Chim. Acta *75*, 169 (1975)
139. Janchen, M., Grunig, G., Bertermann, K.: Anal. Lett. *18*, 1799 (1985)
140. Schindler, J. G., Gulich, M. V.: Fresenuis Z. Anal. Chem. *308*, 434 (1981)
141. Mizutani, F., Sasaki, K., Shimura, Y.: Anal. Chem. *55*, 35 (1983)
142. Clark, L. C., Noyes, L. K., Grooms, T. A., Gleason, C. A.: Clin. Biochem. *17*, 288 (1984)
143. Mascini, M., Moscone, D., Palleschi, G.: Anal. Chim. Acta *157*, 45 (1984)
144. Mizutani, F., Tsuda, K., Karube, I., Suzuki, S., Matsumato, K.: ibid. *118*, 65 (1980)
145. Tsuchida, T., Yoda, K.: Clin. Chem. *29*, 51 (1983)
146. Greibee, R. J., Knoblock, E. C., Koch, T. R.: ibid. *27*, 163 (1981)
147. Nanjo, M., Guilbault, G. G.: Anal. Chim. Acta *75*, 169 (1975)
148. Lubrano, G. J., Guilbault, G. G.: ibid. *97*, 229 (1978)
149. Blaedel, W. J., Jenkins, R. A.: Anal. Chem. *47*, 1337 (1975)
150. Moiroux, J., Elving, P. J.: ibid. *50*, 1056 (1978)

151. Moiroux, J., Elving, P. J.: J. Am. Chem. Soc. *102*, 6533 (1980)
152. Jaegfeldt, H.: J. Electroanal. Chem. *110*, 295 (1980)
153. Carlson, B. W., Miller, L. L., Neta, P., Grodkowski, J.: J. Am. Chem. Soc. *106*, 7233 (1984)
154. Blankespoor, R. L., Miller, L. L.: J. Electroanal. Chem. *171*, 231 (1984)
155. Blaedel, W. J., Engstrom, R. C.: Anal. Chem. *52*, 1691 (1980)
156. Yao, T., Kobayashi, Y., Musha, S.: Anal. Chim. Acta *139*, 363 (1982).
157. Cenas, N., Rozgaile, J., Kulys, J.: Biotech. Bioeng. *26*, 552 (1984)
158. Wallace, T. C., Coughlin, R. W.: Anal. Biochem. *80*, 133 (1977)
159. Bartalits, L., Nagy, G., Pungor, E.: Anal. Lett. *17*, 13 (1984)
160. Luval, J.-M., Bourdillon, C., Moiroux, J.: J. Am. Chem. Soc. *106*, 4701 (1984)
161. Bartalits, L., Nagy, G., Pungor, E.: Clin. Chem. *30*, 1780 (1984)
162. Yao, T., Musha, S.: Anal. Chim. Acta *160*, 141 (1984)
163. Suaud-Chagny, M. R., Gonon, F. G.: Anal. Chem. *58*, 413 (1986)
164. Pan, C. P., Rechnitz, G. A.: Anal. Chim. Acta *160*, 141 (1984)
165. Cheng, F., Christian, G.: Anal. Chem. *49*, 1785 (1977)
166. Cheng, F., Christian, G. D.: Clin. Chem. *24*, 621 (1978)
167. Cheng, F., Christian, G.: Clin. Chim. Acta *91*, 2951 (1979)
168. Kelley, T. A., Christian, G. D.: Analyst *109*, 453 (1984)
169. Williams, D. L., Doig, A. R., Korosi, A.: Anal. Chem. *42*, 118 (1970)
170. Attiyat, A. S., Christian, G. D.: Anal. Chim. Acta *106*, 225 (1979)
171. Smith, M. D., Olson, C. L.: Anal. Chem. *46*, 1544 (1974)
172. Winartasaputra, H., Kuan, S. S., Guibault, G. G.: ibid. *54*, 1987 (1982)
173. Tse, D. C.-S., Kuwana, T.: ibid. *50*, 1315 (1978)
174. Jaegfeldt, H., Torstensson, A. B. C., Johansson, G.: ibid. *53*, 1979 (1981)
175. Jaegfeldt, H., Kuwana, T., Johansson, G.: J. Am. Chem. Soc. *105*, 1805 (1983)
176. Degrand, C., Miller, L. L.: ibid. *102*, 5731 (1980)
177. Huck, H., Schmidt, H.-L.: Angew. Chem. *20*, 402 (1981)
178. Huck, H., Schelter-Graf, A., Danzer, J., Kirch, P., Schmidt, H.-L.: Analyst *109*, 147 (1984)
179. Torstensson, A., Gorton, L.: J. Electroanal. Chem. *130*, 293 (1981)
180. Gorton, L., Torstensson, A., Jaegfeldt, H., Johansson, G.: ibid. *161*, 103 (1984)
181. Appelquist, R., Marko-Varga, G., Gorton, L., Torstensson, A., Johansson, G.: Anal. Chim. Acta *169*, 237 (1985)
182. Huck, H.: Fresnius Z. Anal. Chem. *313*, 548 (1982)
183. Gorton, L., Johansson, G., Torstensson, A.: J. Electroanal. Chem. *196*, 81 (1985)
184. Kulys, J. J.: Enzyme Microb. Technol. *3*, 344 (1981)
185. Albery, W. J., Bartlett, P. N.: J. Chem. Soc. Chem. Commun. 234 (1984)
186. Kulys, J. J.: Biosensors *2*, 3 (1986)
187. Duine, J. A., Frank, J.: Trends in Biochem. Sci. *6*, 278 (1981)
188. Mullen, W. H., Churchouse, S. J., Vadgama, P. M.: Analyst *110*, 925 (1985)
189. Turner, A. P. F., Pickup, J. C.: Biosensors *1*, 85 (1985)
190. Nilsson, H., Akerlund, A., Mosbach, K.: Biochim. Biophys. Acta *320*, 529 (1973)
191. Ruzicka, J., Hansen, E. H., Ghose, A. K., Mottola, H. A.: Anal. Chem. *51*, 199 (1979)
192. Vadgama, P. M., Alberti, K. G. M. M., Covington, A. K.: Anal. Chim. Acta *136*, 403 (1982)
193. Alexander, P. W., Joseph, J. P.: ibid. *131*, 103 (1981)
194. Kulys, J. J., Gureviciene, V. V., Laurinavicius, V. A., Jonuska, A. V.: Biosensors *2*, 35 (1986)
195. Kirstein, D., Kirstein, L., Scheller, F.: ibid. *1*, 117 (1985)
196. Guibault, G. G., Montalvo, J. G.: J. Am. Chem. Soc. *91*, 2164 (1969)
197. Guilbault, G. G., Nagy, G.: Anal. Chem. *45*, 417 (1973)
198. Thanei-Wyss, U., Morf, W. E., Leinemann, P., Stefanac, Z., Mostert, I., Dorig, R., Dohner, R. E., Simon, W.: Mikrochim. Acta *3*, 135 (1983)
199. Yasuda, K., Miyagi, H., Hamada, Y., Tokata, Y.: Analyst *109*, 61 (1984)
200. Anfalt, T., Granelli, A., Jagner, D.: Anal. Lett. *6*, 969 (1973)
201. Johansson, G., Ogren, L.: Anal. Chim. Acta *84*, 23 (1976)
202. Papastathopoulas, D. S., Rechnitz, G. A.: ibid. *79*, 17 (1975)
203. Mascini, M., Guilbault, G. G.: Anal. Chem. *49*, 795 (1977)
204. Guilbault, G. G., Shu, F.: ibid. *44*, 2161 (1972)

205. Winquist, F., Spatz, A., Lundstrom, I., Danielsson, B.: Anal. Chim. Acta *164*, 124 (1984)
206. Thompson, H., Rechnitz, G. A.: Anal. Chem. *46*, 246 (1974)
207. Guilbault, G. G., Chen, S. P., Kuan, S. S.: Anal. Lett. *13*, 1607 (1980)
208. Mascini, M., Palleschi, G.: Anal. Chim. Acta *136*, 69 (1982)
209. Mascini, M., Fortunati, S., Moscone, D., Palleschi, G.: ibid. *171*, 175 (1985)
210. Winquist, F., Lundstrom, I., Danielsson, B.: Anal. Chem. *58*, 145 (1986)
211. Havas, J., Guilbault, G. G.: ibid. *54*, 1991 (1982)
212. Tran, N. D., Romette, J. L., Thomas, D.: Biotech. Bioeng. *25*, 329 (1983)
213. Bradley, C. R., Rechnitz, G. A.: Anal. Chem. *56*, 664 (1984)
214. Heineman, W. R., Halsall, H. B.: ibid. *57*, 1321A (1985)
215. Del Castillo, J., Rodriguez, A., Romero, C. A., Sanchez, V.: Science *153*, 185 (1966)
216. Barfort, P., Arquilla, E. R., Vogelhut, P. O.: ibid. *160*, 1119 (1968)
217. Janata, J.: J. Am. Chem. Soc. *97*, 2914 (1975)
218. Aizawa, M., Kato, S., Suzuki, S.: J. Membr. Sci. *2*, 125 (1977)
219. Aizawa, M., Suzuki, S., Nagamura, Y., Shinohara, R., Ishiguro, I.: Chem. Lett. *1977*, 779 (1977)
220. Meyerhoff, M., Rechnitz, G. A.: Science *195*, 494 (1977)
221. Aizawa, M., Suzuki, S.: J. Solid-Phase Biochem. *4*, 25 (1979)
222. Yamamoto, N., Nagasawa, Y., Shuto, S., Sawai, M., Sudo, T., Tsubomura, H.: Chem. Lett. *1978*, 245 (1978)
223. Yamamoto, N., Nagasawa, Y., Sawai, M., Sudo, T., Tsubomura, H.: J. Immunol. Methods *22*, 309 (1978)
224. Yamamoto, N., Nagasawa, Y., Shuto, S., Tsubomura, H., Sawai, M., Okumura, H.: Clin. Chem. *26*, 1569 (1980)
225. Malmros, M. K.: U. S. Patent #4,334,880 (1980)
226. Collins, S.; Janata, J.: Anal. Chim. Acta. *136*, 93 (1982)
227. Janata, J., Blackburn, G. F.: Ann. N.Y. Acad. Sci. *428*, 286 (1984)
228. Solski, R. L., Rechnitz, G. A.: Science *204*, 1308 (1979)
229. Solski, R. L., Rechnitz, G. A.: Anal. Chim. Acta *123*, 135 (1981)
230. Solski, R. L., Rechnitz, G. A.: U. S. Patent #4,402,819 (1983)
231. Keating, M. Y., Rechnitz, G. A.: Anal. Chem. *56*, 801 (1984)
232. Monroe, D.: ibid. *56*, 921A (1984)
233. Cardosi, M. F., Stanley, C. J., Turner, A. P. F.: Proceedings of the 2nd International Meeting on Chemical Sensors, p. 634, Bordeaux, France, July 7–10, 1986
234. Breyer, B., Radcliff, F. J.: Nature *167*, 79 (1951)
235. Breyer, B., Radcliff, F. J.: Aust. J. Exp. Biol. *31*, 167 (1953)
236. Alexander, P. W., Rechnitz, G. A.: Anal. Chem. *46*, 1253 (1974)
237. Solsky, R. L., Rechnitz, G. A.: Anal. Chim. Acta *99*, 241 (1978)
238. Boitieux, J.-L., Desmet, G., Thomas, D.: Clin. Chem. *25*, 318 (1979)
239. Meyerhoff, M. E., Rechnitz, G. A.: Anal. Biochem. *95*, 483 (1979)
240. Alexander, P. W., Maltra, C.: Anal. Chem. *54*, 68 (1982)
241. Aizawa, M., Morioka, A., Matsuoka, H., Suzuki, S., Nagamura, Y., Shinohara, R., Ishiguro, I.: J. Solid-Phase Biochem. *1*, 319 (1976)
242. Aizawa, M., Morioka, A., Suzuki, S.: Anal. Chim. Acta *115*, 61 (1980)
243. Itagaki, H., Hakoda, Y., Suzuki, Y., Haga, M.: Chem. Pharm. Bull. *31*, 1283 (1983)
244. Karube, I., Matsunaga, T., Suzuki, S., Asano, T., Itoh, S.: J. Biotechnology *1*, 279 (1984)
245. Ikariyama, Y., Furuki, Aizawa, M.: Anal. Chem. *57*, 496 (1985)
246. Yuan, C.-L., Kuan, S. S., Guilbault, G. G.: ibid. *53*, 1906 (1981)
247. Robinson, G. A., Cole, V. M., Rattle, S. J., Forrest, G. C.: Biosensors *2*, 45 (1986)
248. Weber, S. G., Purdy, W. C.: Anal. Lett. *12*, 1 (1979)
249. Di Gleria, K., Hill, H. A. O., McNeil, C. J., Green, M.: Anal. Chem. *58*, 1203 (1986)
250. Wehmeyer, K. R., Halsall, H. B., Heineman, W. R.: Clin. Chem. *28*, 1968 (1982)
251. Heineman, W. R., Anderson, C. W., Halsall, H. B.: Science *204*, 865 (1979)
252. D'Orazio, P., Rechnitz, G. A.: Anal. Chim. Acta *109*, 25 (1979)
253. Gebauer, C. R., Rechnitz, G. A.: Anal. Biochem. *103*, 280 (1980)
254. Shiba, K., Watanabe, T., Umezawa, Y., Fujiwara, S.: Chem. Lett. 155 (1980)
255. Shiba, K., Umezawa, Y., Watanabe, T., Ogawa, S., Fujiwara, S.: Anal. Chem. *52*, 1610 (1980)
256. Haga, M., Itagaki, H., Sugawara, S., Okano, T.: Biochem. Biophys. Res. Comm. *95*, 187 (1980)

257. Brontman, S. B., Meyerhoff, M. E.: Anal. Chim. Acta *162*, 363 (1984)
258. Gebauer, C. R., Rechnitz, G. A.: Anal. Biochem. *124*, 338 (1982)
259. Eggers, H. M., Halsall, H. B., Heineman, W. R.: Clin. Chem. *28*, 1848 (1982)
260. Broyles, C. A., Rechnitz, G. A.: Anal. Chem. *58*, 1241 (1986)
261. Fonong, T., Rechnitz, G. A.: ibid. *56*, 2586 (1984)
262. Ngo, T. T., Bovaird, J. H., Lenhoff, H. M.: Appl. Biochem. Biotech. *11*, 63 (1985)
263. Oberhardt, B. J., Wotherspoon, N.: U. S. Patent 4,280,815 (1981)
164. Olsson, T., Thore, A.: Chemiluminescence and Its Use in Immunoassay in: Immunoassays for the 80's (eds.) Voller, A., Bartlett, A., Bidwell, D. p. 113, Baltimore, Maryland, University Park Press 1981
the 80's (eds. Voller, A., Bartlett, A., Bidwell, D.) p. 113, New York, MTP Press 1981
265. Ikariyama, Y., Kunoh, H., Aizawa, M.: Biochem. Biophys. Res. Comm. *128*, 987 (1985)
266. Ege, D., Becker, W. G., Bard, A. J.: Anal. Chem. *56*, 2413 (1984)

Photoelectrochemical Solar Energy Conversion

Rüdiger Memming

Philips GmbH Forschungslaboratorium Hamburg
Vogt-Kölln-Str. 30, D-2000 Hamburg 54, FRG

Table of Contents

In the present paper the progress in the field of solar energy conversion for the production of electricity and storable chemical fuels during the last decade is reviewed. The current-potential behavior of regenerative photovoltaic cells are derived and related to charge transfer processes at the solid/liquid interface. Various cells are critically analyzed in view of their stability and conversion efficiency. A number of factors limiting the photovoltage are discussed in terms of a stabilization mechanism, trapping of minority carriers at the interface and the forward dark current. Concerning the production of chemical fuels the photocleavage of water and hydrogen sulfide, the reduction of carbon dioxide and the formation of ammonia is evaluated. The main emphasis is laid here on catalytic processes at semiconductor electrodes and particles. The principle function of catalysts being deposited on extended electrodes and particles are discussed in detail.

1 Introduction

The great energy consumption, limited recources of traditional fuels and environmental problems have lead to intensive research on the conversion of solar energy during the last fifteen years. Conversion into electrical energy has been realized in technical devices consisting of pn-junction photovoltaic cells. Efficiencies of up to 20% have been obtained with single crystal devices and around 9% with polycrystalline or amorphous layers.

During the last decade an ever increasing interest has been directed toward photochemical and photoelectrochemical systems, the latter being applicable for solar energy conversion into electrical or chemical energy. Some of these processes are similar to those occuring in nature, i.e. in photosynthesis of plants. In nature, however, the efficiency is hardly above 1%. In addition biological systems exhibit a rather low lifetime which is not critical because plants regenerate themselves every year. In the case of technical applications, however, an efficiency of about 10% and a stability for at least 10–15 years are required which is a great challenge for research on photoelectrochemical systems.

Hundreds of papers, summarized partly in review articles [1-13], were published in this field. In the present paper not only the state of art is given but also the essential effects and problems are described and analyzed in detail. Concerning the fundamentals of semiconductor electrochemistry which are not introduced here, it must be referred to corresponding review articles[14-18].

2 Photovoltaic Cells

As it has been described in various other review articles before, the conversion efficiencies of photovoltaic cells depend on the band gap of the semiconductor used in these systems[2,7,19]. The maximum efficiency is expected for a bandgap around $E_G = 1.3\,eV$. Theoretically, efficiencies up to 30% seem to be possible[19]. Experimental values of 20% as obtained with single crystal solid state devices have been reported[20]. Since the basic properties are identical for solid/solid junctions and for solid/liquid junctions the same conditions for high efficiencies are valid. Before discussing special problems of electrochemical solar cells the limiting factors in solid photovoltaic cells will be described first.

2.1 Solid State Photovoltaic Cells

In principle there are two types of solid state devices: (i) pn-photocells and (ii) Schottky type cells. The first one consists simply of a pn-junction whereas the other of a semiconductor-metal junction. The energy schemes of these cells are given in Fig. 1a and b. The current-potential dependence of both types of cells is given by (see e.g.[21]):

(i) $$i = i_0 \left(\exp \frac{eU}{kT} - 1 \right) - i_{ph} \tag{1}$$

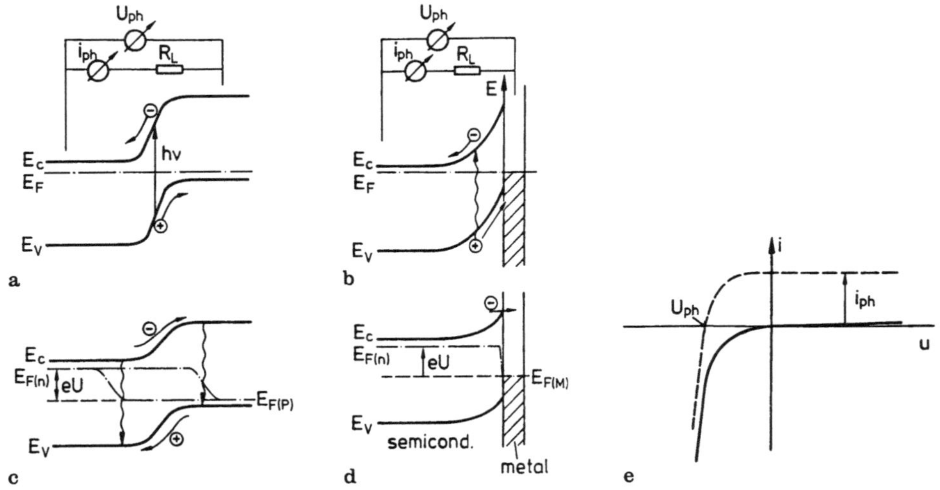

Fi.g 1 a—e. Charge transfer processes at pn-junctions (left side) and semiconductor-metal Schottky junctions (right side).
a) and **b)** photoeffects; **c)** and **d)** charge transfer under forward bias in the dark; **e)** current-potential behavior.
E_c = conduction band, E_V = valence band, E_F = Fermi level; E_F = Termi level

in which i_0 is the saturation current, U the externally applied voltage and i_{ph} the photocurrent. A typical current-voltage behavior in the dark (solid line) and under illumination (dashed curve) is shown in Fig. 1e. The light excitation leads to the formation of electron-hole pairs which are seperated by the electric field across the space charge layer as indicated in Figs. 1a and b. The corresponding photocurrent in the reverse direction is proportional to the light intensity. The photopotential U_{ph} as measured under open circuit conditions, occurs at that potential at which the photocurrent is equal to the corresponding dark current, i.e. where the total current is zero under a certain illumination (see Fig. 1c). According to Eq. 1 one obtains then:

$$i = 0$$

$$U_{ph} = \frac{kT}{e} \ln \left[\frac{i_{ph}}{i_0} + 1 \right] \tag{2}$$

The photovoltage is esentially determined by the ratio of the photo- and saturation current. Since i_0 occurs as a pre-exponential factor in Eq. 1 it determines also the dark current. Actually this is the main reason that it limits the photovoltage via Eq. 2. The value of i_0 depends on the mechanism of charge transfer at the interface under forward bias and is normally different for a pn-junction and a metal-semiconductor contact. In the first case electrons are injected into the p-region and holes into the n-region. These minority carriers recombine somewhere in the bulk as illustrated in Fig. 1c. In such a minority carrier device the forward current is essentially determined by the recombination flow. The saturation current is then given by the Shockley equation (see e.g.[21]), i.e.

$$i_0 = n_i^2 e \left(\frac{D_p}{N_D L_p} + \frac{D_n}{N_A L_n}\right) \quad \text{(minority carrier device)} \tag{3}$$

in which D_p and D_n are the diffusion constants of electrons and holes, L_p and L_n the diffusion lengths of electrons and holes, respectively, and N_D and N_A the corresponding donor and acceptor densities on both sides of the junction.

The forward current at a semiconductor-metal junction is mainly determined by a majority carrier transfer i.e. electrons for n-type, as illustrated in Fig. 1d. In this majority carrier device the socalled thermionic emission model is applied[21] according to which all electrons reaching the surface are transferred to the metal. In this case we have:

$$i_0 = A \frac{m^*}{m_e} T^2 \exp\left(\frac{-eU_B}{kT}\right) \quad \text{(majority carrier device)} \tag{4}$$

in which eU_B is the barrier height of the junction (see Fig. 1b), m^* the effective mass and A is given by

$$A = \frac{4\pi m_e kT}{h^3} = 120 \frac{A}{cm^2} \tag{5}$$

(m_e = electron mass, h = Planck constant)

According to Eq. 4 i_0 is only determined by the barrier height at the semiconductor-metal junction whereas it depends on doping and diffusion length for a pn-junction (see Eq. 3). Taking GaAs (Eg = 1.4 eV) as an example of a doping level of N_D = or $N_A = 10^{17}\ cm^{-3}$ and a typical value for the diffusion length ($L_n = L_p \sim 10^{-4}$ cm) one obtains $i_0 \approx 10^{-18}\ A\ cm^{-2}$ for a corresponding pn-junction ($n_i^2 = 10^{13}\ cm^{-6}$; $D_p = D_n = 10\ cm^2/s$). In the case of a GaAs/metal junction the corresponding value is $i_0 \approx 10^{-17}$ assuming a barrier height of $eU_B = 1.3$ eV and $m^* = 0.07$.

According to this estimate the i_0-values are identical within one order of magnitude. Using a photocurrent of $i_{ph} \sim 2 \cdot 10^{-2}\ A\ cm^2$ a value being typical for sunlight irradiation (AMI) the corresponding photovoltage can be calculated by using Eq. 2. One ontains $U_{ph} \sim 0.9$ Volt; a value which has been verified experimentally with pn-solar cells. In the case of Schottky junctions, however, much smaller values (i.e. $U_{ph} \leqq 0.4$ V) usually the barrier height did not exceed $eU_B = 0.85$ eV[21] so that $i_0 \geqq 10^{-9}\ A\ cm^{-2}$. Only very recently Schottky junctions of GaAs/Au of a barrier height of $eU_B = 1.33$ eV were made[22].

Another important feature of a solar cell is the fill factor. The output power reaches a maximum for a given light intensity and load, i.e.

$$P_{max} = i_{ph(max)} \cdot U_{ph(max)} \tag{6}$$

(see also Fig. 1e).

The fill factor is defined by

$$FF = \frac{i_{ph(max)} \cdot U_{ph(max)}}{i_{ph} \cdot U_{ph}} \tag{7}$$

It reaches a value of about 0.75 for an ideal junction behavior. In practice, however, always lower values have been obtained. This is mainly caused by a deviation of the slope of the current-potential dependence under forward bias, i.e. the i — U dependence is not given by Eq. 1 but is mostly described by

$$i = i_0 \left(\exp \frac{eU}{nkT} - 1 \right) - i_{ph} \,. \tag{8}$$

This relation differs from that given by Eq. 1 in so far as an ideality factor n is introduced in the exponential term. This factor may have different origins[21], which will not be discussed here.

2.2 Electrochemical Photovoltaic Cells

2.2.1 Mechanism of Regenerative Cells

Similar photovoltaic cells as those described above can be made with semiconductor/liquid junctions. The basic function of such a cell is illustrated in terms of an energy scheme in Fig. 2. The system consists of an n-type semiconductor and an inert metal

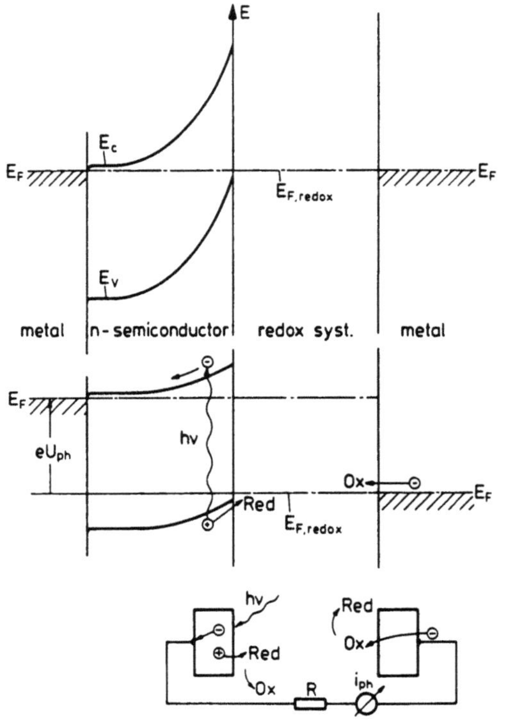

Fig. 2. Schematic energy diagram of a semiconductor/redox system/metal regenerative photovoltaic cell

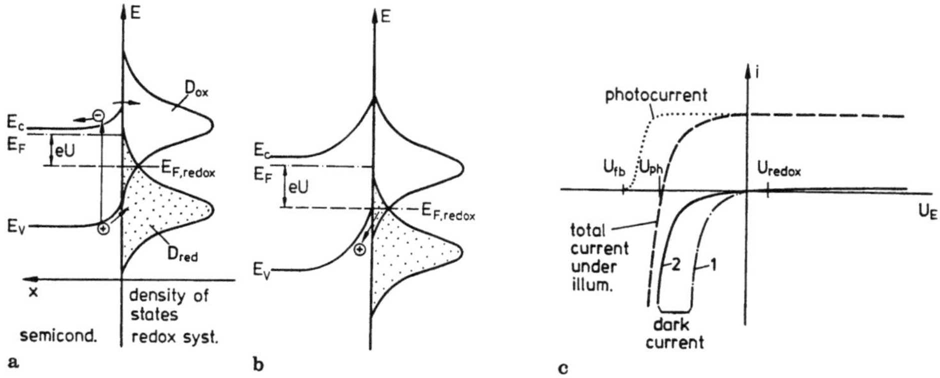

Fig. 3a—c. Charge transfer processes at semiconductor-electrolyte interface **a)** and **b)**: under forward bias.
a) electron transfer via conduction band; **b)** injection of minority carriers (holes); **c)** current-potential behavior.
Curve 1 and 2 correspond to two different $i_{c,o}$-values (see. Eq. 19b), the total current is the sum of photocurrent and dark current (curve 2)

or carbon counter electrode both being in contact to an electrolyte containing a suitable reversible redox system. At equilibrium the electrochemical potential (Fermi level in the solid) is constant throughout the whole system. Upon light excitation the holes move toward the surface where they are consumed for the oxidation of the reduced species (red) of the redox system whereas the electrons move to the ohmic backside contact. Under short circuit conditions or under a certain load the electrons reach the counter electrode where they are used for the reduction of the oxidized species (ox) of the redox system. Accordingly we have

$$\text{red} + h^+ \rightarrow \text{ox} \quad \text{(at n-type electrode)} \tag{9a}$$
$$\text{ox} + e^- \rightarrow \text{red} \quad \text{(at counter electrode)}. \tag{9b}$$

Assuming that no other side reactions occur at the semiconductor electrode this is a cell which operates under completely regenerative conditions.

In order to obtain a large photovoltage a redox couple of a relatively positive standard potential should be selected in order to have a high band bending at equilibrium. The position of the energy bands at the surface are usually determined by the solvent which can easily be determined by capacity measurements (see e.g.[14, 15]). Therefore, it is easy to choose a corresponding redox couple. The question arises, however, which factors limit the photovoltage under sun light conditions. The basic effects are illustrated by a more detailed energy scheme of the semiconductor-electrolyte interface (Fig. 3a). In comparison to Fig. 2 also energy states of the redox system are shown, the occupied states (density D_{red}) of which are located below the Fermi level and the empty states (density D_{ox}) above the Fermi level. Such a semiconductor/liquid interface is similar to a semiconductor/metal interface in various aspects. Differences occur only in the distribution of energy states which is nearly continuous in a metal whereas it is limited to a certain energy range in a redox system. The density of occupied states is given by[14)]

$$D_{red}(E) = D_0 \exp - \left[\frac{(E - E_{F, redox} - \lambda)^2}{4kT\lambda}\right] \tag{10 a}$$

and of the empty states

$$D_{ox}(E) = D_0 \exp - \left[\frac{(E - E_{F, redox} + \lambda)^2}{4kT\lambda}\right] \tag{10 b}$$

in which D_0 is a normalizing factor and λ the reorientation energy. The half-width of the distribution curves is given by

$$\Delta E_{1/2} = 0.53\lambda^{1/2}\ \mathrm{eV}\ . \tag{11}$$

Since the reorientationnergy λ varies in the range of 0.5–2 eV[15)] the half width can be in the order of the bandgap of the semiconductor. Assuming that in the dark the electron transfer occurs entirely via the conduction band (majority carrier device) the current-potential dependence can be derived as follows:

The cathodic partial current is given by

$$\overleftarrow{i}_c = \overleftarrow{k}_c c_{ox} n_s = -\overleftarrow{k}_c c_{ox} n_0 \exp\left(-\frac{eU_{sc}}{kT}\right) \tag{12}$$

in which c_{ox} is the concentration of the oxidized species, n_s and n_o the electron density at the surface and in the bulk, respectively, U_{sc} the potential across the space charge layer and

$$k_c^- = k_{c,o}^- D_{ox}(E_c) \tag{13}$$

The anodic partial current is determined by

$$i_c^+ = k_c^+ c_{red} N_c \tag{14}$$

in which c_{red} is the concentration of the reduced species and N_c the density of energy states in the conduction band, whereas k_c^+ is given by

$$k_c^+ = k_{c,o}^+ D_{red}\ . \tag{15}$$

At equilibrium, i.e. at the redox potential, we have

$$\overleftarrow{i}_c = \overrightarrow{i}_c = i_{c,o} = \overleftarrow{k}_c c_{ox} n_0 \exp - \left(\frac{eU_{sc}^0}{kT}\right). \tag{16}$$

The total current is given by

$$i = i_c^+ - i_c^-\ . \tag{17}$$

Inserting (12) and (14) and using (16) one obtains for the dark current

$$i = -i_{c,o}\left[\exp\left(-\frac{eU}{kT}\right) - 1\right] \tag{18}$$

in which

$$U = U_{sc} - U_{sc}^{0} \tag{19a}$$

and in the presence of illumination

$$i = -i_{c,o}\left[\exp\left(-\frac{eU}{kT}\right) - 1\right] + i_{ph}\,. \tag{19 b}$$

Similarly as in solid state devices the photovoltage U_{ph} is the given by ($i = 0$):

$$U_{ph} = \frac{-kT}{e} \ln\left(\frac{i_{ph}}{i_{c,o}} + 1\right). \tag{20}$$

Again the photovoltage is mainly determined by $i_{c,o}$. However, the latter depends on the rate constant k^- and on the concentration c_{oc} (see Eq. 16) i.e. the photovoltage can be increased by using a lower concentration of the oxidized species as illustrated by current-potential curves in Fig. 3c[23]. There is a limit, however, as the current should not be diffusion limited. Accordingly, higher photovoltages are possible than for semiconductor-metal junctions. It should be emphasized that the current equation was derived on the basis of an interfacial kinetic model. At sufficiently high concentrations the dark current should not be determined by this kinetics but entirely by the transport of electrons through the space charge layer, i.e. the thermionic emission model should be valid as derived for semiconductor-metal junctions and the corresponding i_0 is then given by Eq. 4.

In the model presented above the forward dark current corresponds to an electron transfer via the conduction band. Using, however, a redox couple of a relatively positive standard potential the empty states of the redox system occur rather close to the valence band and the cathodic current could be due to an electron transfer via the valence band as illustrated in Fig. 3b. In this case one still obtains the same i — U characteristic but the saturation current is now given by

$$i_{v,o} = k_v^- c_{ox} N_v \tag{21}$$

in which N_v is the density of states in the valence band. In the dark holes are injected under forward bias which recombine with electrons somewhere in the semicondoctor. Here also $i_{v,o}$ is concentration dependent. At high concentrations the current is not determined by the interfacial kinetics but by the recombination of the injected minority carriers (Shockley model), so that $i_{v,o}$ is given by (Eq. 3).

In conclusion it should be mentioned that the same type of effects are possible for p-type electrodes. In this case an anodic dark current occurs whereas the photocurrent corresponds to an electron transfer via the conduction band (cathodic photocurrent).

2.2.2 Analysis of Various Systems

During the last decade quite a large number of systems have been proposed and investigated. Besides various semiconductors also aqueous and nonaqueous liquids have been used. Severe problems arise mainly in aqueous solutions because other reactions than charge transfer to redox systems can occur, namely O_2- and H_2- evolution and especially anodic dissolution. Concerning the latter problem there are some stable semiconductor electrodes available, however, all of them exhibit a very large bandgap ($\geqq$ 3 eV). Although all other semiconductors undergo corrosion a number of systems have been found in which the anodic decomposition could be sufficiently suppressed in the presence of a suitable redox system. Corresponding systems are given in Table 1. Here only those are listed, which exhibit or promise a high conversion efficiency besides good stability. It is interesting to note that stable cells with n-type electrodes in aqueous solutions were only fabricated by using S^{2-}/S_n^{2-} (or Se^{2-}/Se_2^{2-}) and I^-/I_3^- as redox systems. The latter system may be favorable concerning stabilization because iodine adsorbs rather strongly on surfaces. In the case of chalcogenide electrodes the other system is advantageous because here any elemental sulfur or selenium formed as corrosion products is dissolved as polysulfide or polyselenide. Details on the stabilization problem will be discussed seperately in the next chapter.

Most authors have characterized the cell performance by a power plot, i.e. i_{ph} vs. U_{ph}, and not by a complete i — U characterist in the dark and under illumination. Two examples are given in Fig. 4 which are typical for many investigations. The shape of these curves and consequently the fill factor (defined by (Eq. 7)) depends on the surface treatment, such as etching or deposition of metal atoms and on the composition of the solution. The origin of a deviation from an ideal i_{ph} — U_{ph} characteristic may have different reasons such as a formation of a surface layer, surface recombination and reaction kinetics. Especially the group of the Weizmann Institute has investigated the performance of chaleogenide electrodes in (S^{2-}/S_n^{2-})-solutions in detail ([37] and literature cited there). They did not only find an influence of etching and pH of the solution[38] but also of the cations such as various alkali as illustrated in Fig. 4a.

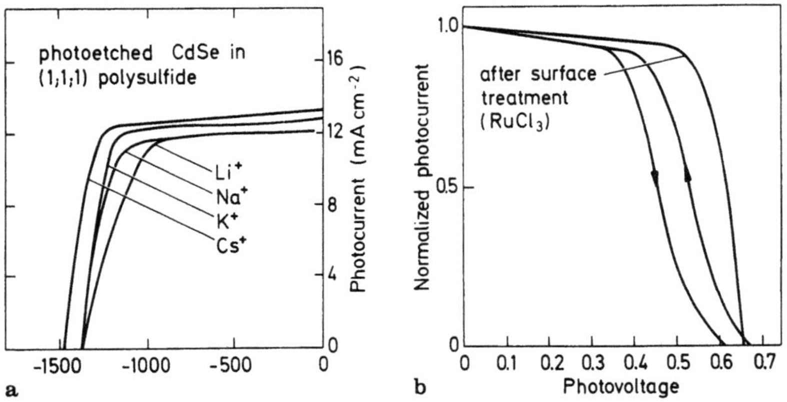

Fig. 4a and b. Power plot for **a)** CdSe/(S^{2-}/S_n^{2-})[39]; **b)** GaAs/(Se^{2-}/Se_2^{2-})[27]

Table 1. Photovoltage U_{ph}, photocurrent i_{ph} fill factor FF and efficiency η of regenerative electrochemical solar cells

No	Cell	E_g [eV]	U_{fb}-U^0_{redox} [V]	Solvent	U_{ph} [V]	i_{ph} [m A cm^{2-}]	FF	η %	References
1.	n-CdSe/($S^{2-}/^{2-}_{n}$)	1.7	0.8	NaOH	0.75	12	—	8	[24]
2.	n-Cd(Se, Te)/(S^{2-}/S^{2-}_n)	1.7	0.8	H_2O	0.78	22	0.65	12.7	[25]
3.	n-GaAs/(Se^{2-}/Se^{2-}_n)	1.4	0.7	NaOH	0.65	20	—	12	[26, 27]
4.	n-CdS/(I^-/I_3^-)	2.5	1.0	CH_3CN	0.95	0.03	—	9.5	[28]
5.	n-$CuInSe_2$/(I^-/I_3^-)	1.01	—	H_2O	0.64	21	—	9.7	[29, 30]
6.	n-$MoSe_2$/(I^-/I_3^-)	1.1	0.5	H_2O	0.55	9	—	—	[31]
7.	n-WSe_2/(I^-/I_3^-)	1.2	0.65	H_2O	0.63	28	—	> 14	[32]
8.	n-Si/(Br^-/Br_2)[(a)]	1.1	1.0	H_2O	0.68	22	—	—	[33]
9.	n-Si/(Fc^+/Fc)[(b)]	1.1	—	CH_3OH	0.67	20	> 0.7[(c)]	> 10[(c)]	[34]
10.	n-GaAs/(Fc^+/Fc)[(b)]	1.4	—	CH_3CN	0.7	20	—	11	[35]
11.	p-InP/(V^{3+}/V^{2+})	1.3	0.9	HCL	0.65	25	0.64	11.5	[36]

[a] discussed in Sect. 2.3; [b] Fc = ferrocene derivative; [c] estimated

The authors relate this effect to ion pairing for strongly hydrated cations such as Li^+ which results in a decreased activity of the active (poly) sulfide at the electrode. Using Cs^+ instead of Li^+ not only the electrochemical kinetics but also the stability would be considerably improved[39].

Concerning $CuInX_2$ electrodes (X = S or Se) it was interestingly found, that these electrodes show a much better stability in polysulfide solution than CdX_2 but the fill factor was rather poor in corresponding solar cells[37]. On the other hand in iodide solutions a high fill factor and a good conversion efficiency was found but $CuInSe_2$ is unstable in I_3^-[29,40]. Surprisingly, the stability could be extremely increased by adding Cu^+-ions to the electrolyte [29,40]. It has been shown that the stabilization is caused by the formation of a passivating layer consisting of p-type $CuISe_3$-Se^0[41]. Accordingly, a pn-heterojunction is formed which is actually the active part of the solar cell, i.e. we have here a solid state solar cell in contact with an electrolyte.

Another promising n-type material seems to be FeS_2 (pyrite) which has a relatively low bandgap of $E_g = 0.95$ eV. It is reported that this material shows a very good stability in I^-/I_3^- solutions even under strong illumination[42]. For instance a charge of more than 6×10^5 Coulomb cm^{-2} were passed through the cell without any evidence of corrosion. This material is very cheap and it exhibits a very high absorption coefficient ($\alpha > 6 \times 10^5$ cm^{-1} for $E_{ph} > 1.3$ eV) which is advantageous for fabricating thin film devices[43].

In several cases also the deposition of metal atoms at special sites improved the $i_{ph} - U_{ph}$ characterists considerably as shown for Ru on GaAs[27] (Fig. 4b) and Cu on CdSe[44]. This effect has been interpreted by assuming a decrease of surface recombination due to strong interaction between the Ru and the GaAs-surface atoms resulting in a splitting of surface states into two new states which are not active anymore[45]. It is an open question, however, whether the metal being deposited on the surface influences catalytically the reaction rate. More details on the influence of surface recombination and catalytic effects will be given in seperate chapters (see below).

Very little can be concluded form the experimental results with respect to theoretical current-potential characteristics in the dark and under illumination derived in the previous chapter. Since in none of these cases a complete i-U-curve was published or

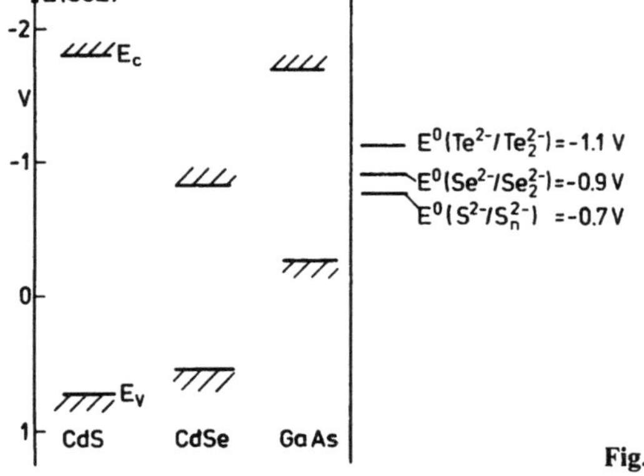

Fig. 5. Position of energy bands at the interface

analyzed there are no data available on the saturation current or on the slope of i — U curve in forward direction and only very little can be concluded with respect to the rate determining steps. It is interesting to note that in most cells listed in table I (at least for No. 1–7) the distance between flatband potential and standard redox potential, i.e. $U_{fb} - U^0_{redox}$, and consequently between conduction band E^s_c and U^0_{redox} is about half of the bandgap as also illustrated for some examples in Fig. 5. Since in these cases the empty energy states of the redox couple occur rather close to the conduction band the forward dark current should correspond to an electron transfer from the conduction band to the acceptor in the electrolyte (compare with Fig. 3). It is also remarkable that the photovoltage measured with cells No. 1–7 (Table 1) at arbitrary photocurrent densities are nearly identical to $(U_{fb} - U^0_{redox})$; i.e. the energy bands become flat at these light intensities. From these results it has to be concluded that the corresponding $i_{c,o}$-values must be extremely small (see Eq. 20). Accordingly, the majority carrier current — electron transfer via the conduction band — is controlled by the kinetics at the interface. In summary we have:

$$X + e^- \rightarrow X^- \quad \text{(for No. 1–7 in table I)} \tag{22}$$
$$(X = S\ ;\ \ Se \text{ or } I_3^-)$$
$$X^- + h^+ \rightarrow X \tag{23}$$

in which the holes are created by light excitation. There may be surface states involved in the hole transfer process which will be discussed in the next chapters.

In this context the Si/ferrocene cell operated with CH_3OH as a liquid is of special interest because its function and the rate determining steps have been analyzed in more detail[34, 35]. Using Si as an electrode in CH_3OH or CH_3CN the solvent has to be extremely H_2O-free in order to avoid any growth of a thicker oxide layer which would block a charge transfer across the interface. It was found that the photovoltage rises linearly with increasing redox potential ($\Delta U_{ph} = \Delta U^0_{redox}$) in a range of about 0.55 V, the redox potential being varied by using different derivatives of ferrocene. The highest photovoltage was 0.67 V at $i_{ph} = 20$ m A cm^{-2} (bandgap 1.1 eV). In addition it was shown that the forward current is a valence band process, i.e.

$$Fc^+ \rightarrow Fc + h^+ \quad \text{(at Si-electrode)} \tag{24}$$

at least when dimethyl ferrocene ($U^0_{redox} = 0.15$ V (SCE)) was used. Interestingly, this forward dark current was not controlled by the interface kinetics but entirely by diffusion and recombination of holes injected into the valence band. Accordingly the photovoltage is determined by Eq. 2 and the i_0-value by Eq. 3. This was checked by using differently doped Si-electrodes having different diffusion lengths of holes[34]. Quantitative agreement was found between experimental and theoretical photovoltages, the latter being calculated by using Eqs. 2 and 3. The result that the forward current corresponds to a valence band process, however, is rather surprising because of the following reasons:

Since the current is entirely controlled by the recombination of injected holes it has to be concluded that the interface kinetics must be fast. This is only possible, if the density of empty states of the ferrocence redox couple is sufficiently high at the edge of the valence band (Fig. 3). Unfortunately the authors did not determine the

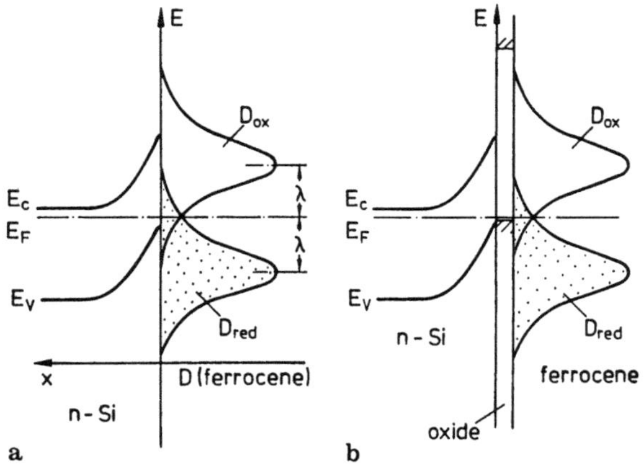

Fig. 6a and b. Energy scheme of n-Si/ferrocene in methanol

flatband potential so that one can only estimate the relative positions of energy states on both sides of the interface as shown in Fig. 6a. Since ferrocence is a nice outer sphere redox system the rearrangment energy can be estimated to be around $\lambda = 0.5 - 0.7$ eV[34]. According to Fig. 6a it is quite obvious, however, that the density of empty states of the ferrocene couple must be much higher at the conduction band than around the valence band so that one would expect preferably a conduction band and not a valence band process! Such a discrepancy which has recently also been observed with other semiconductor electrodes such as p-GaAs and p-InP[23,46], can only be solved by assuming an additional selectivity process as illustrated in Fig. 6b. Here it is assumed that a thin oxide layer is present on the Si-surface. Provided that occupied states exist within the oxide, holes can easily be injected. An electron transfer via the conduction band is inhibited if empty states in the oxide are located far above the conduction band of Si[23]. Such surface layers with selective properties may also play a role in other solar cells. Further investigations are required for testing this model.

In the case of layer compounds as electrode materials the kinetics of charge transfer were also studied in some detail taking into account surface recombination which plays an important role here[47]. In the presence of suitable redox systems some materials show very little corrosion[48]. This is due to the morphology of the crystal surfaces and it is generally assumed that corrosion occurs only at steps of different crystal planes[49]. Accordingly, it is not surprising that the highest efficiencies were obtained with some of these materials (Table 1)[32]. The steps also play an important role in the fill factor as determined by surface recombination measurements[50].

Finally cells containing a p-type semiconductor electrode should be mentioned. In principle the application of p-type electrodes would be even more favorable because electrons created by light excitation are transferred from the conduction band to the redox system. Stability problems are less severe because most semiconductors do not show cathodic decomposition (see e.g. earlier review article[15]. However, there is only one system, p-InP/(V^{2+}/V^{3-}), with which a reasonable efficiency was obtained (Table 1)[36]. There are mainly two reasons why p-electrodes were not widely used: (i) not many materials are available from which p-type electrodes can be made; (ii) most p-type electrodes exhibit a rather high overvoltage for the onset of the photocurrent with respect to the flatband potential. This is due to strong surface recombination[7], trapping of holes[51], surface film formation[52], and in the case of layer compounds to intercalation effects. Several attempts have been made to overcome this problem, however, without a real breakthrough.

2.2.3 Stabilization of Semiconductor Electrodes

As already mentioned before the stabilization of semiconductor electrodes contacting an aqueous electrolyte is the most severe problem in regenerative solar system and even more in photoelectrolysis cells. Therefore, it is important to get more insight into this problem. In this context it is interesting to note that in all stable systems the standard potential of the redox system occurs around or above the midgap of the corresponding semiconductor (see Table 1 and Fig. 5), i.e. $(U_{fb} - U^0_{redox}) \leqq 0.5\,Eg$. This result implies that surface states located in the middle of the bandgap may be involved in the charge transfer process as indicated in Fig. 7a. Since such a process can lead to good stabilization it has been suggested that surface states occur as intermediates in the anodic dissolution process[53] according to the reaction

$$S + h^+ \rightarrow \dot{S}^+ \qquad (25)$$

$$\dot{S}^+ + (m-1)\,h^+ + mX^- \rightarrow SX_m \quad \text{anodic dissol.} \qquad (26)$$

$$\dot{S}^+ + Red \rightarrow S + Ox \quad \text{redox process} \qquad (27)$$

in which S represents a semiconductor surface molecule and $\dot{S}^+$ a surface radical. These reactions are illustrated in an atomistic picture in Fig. 7b[7]. Such a radical can act as a surface state.

Concerning the competition between redox process and anodic decomposition one can define a stability factor s by

$$s = \frac{i_{ox}}{I_{ph}} \qquad (28)$$

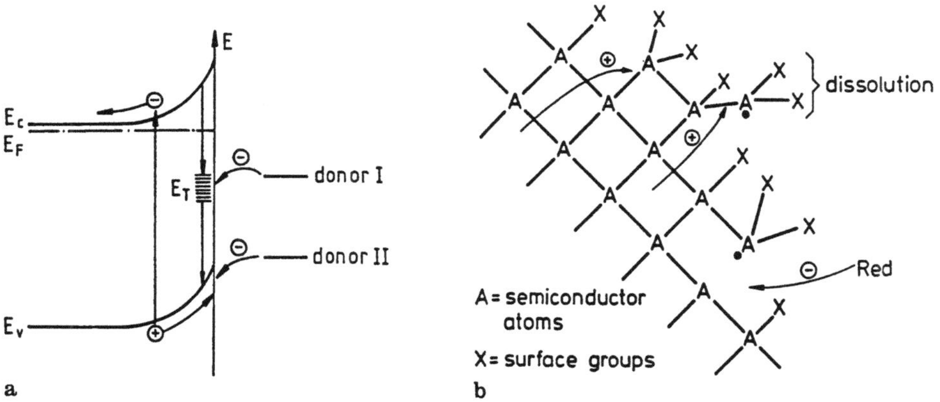

Fig. 7a and b. Stabilization by surface state reactions. **a)** energy scheme; **b)** molecular picture in which i_{ox} represents the current due to the oxidation of the redox system whereas the photocurrent i_{ph} is given by

$$i_{ph} = i_{ox} + i_{corr}$$

in which i_{corr} is the corrosion current. The first quantitative measurments have already shown that s varies with pH indicating that not only the relative position of bands are important but also other kinetic factors are involved[53]. Later on it was found, especially with n-GaAs and n-GaP, that the stabilization factor depends strongly on light intensity and concentration of the redox system[54-57]. Especially Gomes and his group has studied these phenomena more quantitatively[56]. There is only one system, n-WSe_2/($[Fe(CN)_6]^{4-}$), for which it was reported that s is independent of light intensity[58]. These results indicate that the reaction scheme given by reactions[25-27] is too simple. Various detailed models have been derived[58,59], some of them being rather complicated. Recently, a relatively simple reaction scheme has been postulated[60] which includes all essential kinetic aspects already introduced by other authors. The reaction scheme is given by the following reactions (Eq. 30–32):

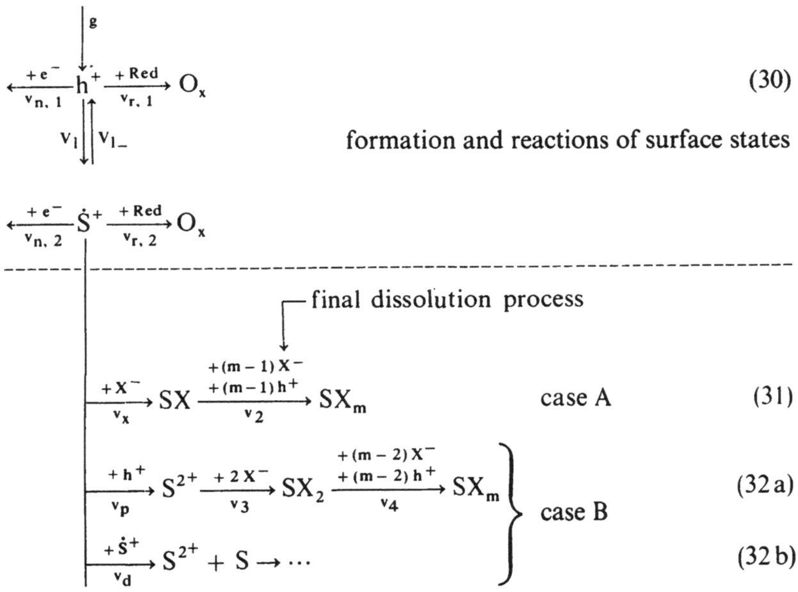

In the first part of the reaction scheme (Eq. 30) the generation (g) of holes (h^+) in the valence band and of the surface radical $\dot{S}^+$ is described. The holes can also be consumed by recombination with electrons (rate $v_{n,1}$) or by direct hole transfer to the redox system ($v_{r,1}$). The surface radical $\dot{S}^+$ can react with an electron from the conduction band ($v_{n,2}$) or with the redox system ($v_{r,2}$), processes by which the radical disappears. Accordingly, the original bond is repaired as illustrated in Fig. 7b.

The lower part of the reaction scheme describes three possibilities for further dissolution reactions of the radical (Eq. 31 and 32). Only the first step in this sequence is essential. It is important to note here that in one path (case A) the first step is a pure chemical reaction characterized by the rate v_x, whereas in the others (case B) a hole transfer is involved. The different kinetics yield the following results[60]:

case A

$$\frac{S}{1-S} = \alpha c_{red} + \beta c_{red}^2 \tag{33}$$

with $$\alpha = \frac{k_1 k_{r,2} + k_{r,1}(k_x c_x + k_n{}' {}_2 n_s + k_{1-})}{m k_x c_x k_1}$$

$$\beta = \frac{k_{r,1}, k_{r,2}}{m k_x c_x k_1}$$

case B

$$\frac{s}{1-s} = \frac{k_{r,1} c_{red}}{m k_1} + \frac{f(c_{red})}{g} \tag{34}$$

with $$f(c_{red}) = \frac{(k_1 + k_{r,1} c_{red})^2 \, k_{r,2} c_{red}}{m k_1 k_p}$$

(assuming $n_s = 0$; $k_{1-} = 0$)

According to case A (see Eq. 33) the stability factor is independent of light intensity as found with WSe_2[58]. In this case, however, one cannot distinguish experimentally whether a direct hole transfer from the valence band to the redox system occurs or whether surface states (radicals) are involved. In case B the stability decreases with increasing hole generation, i.e. light intensity (Eq. 34) could be verified experimentally in some cases[60]. Recently, a further model has been published in which also intermediates originating from further reactions of SX_2 are involved[61]. There are also other indications for the participation of intermediates in surface reactions, such as for instance in the formation of sulfate as a corrosion product of CdS in the presence of oxygen[62] and in etching processes at GaAs[63]. Certainly these intermediates are important in the stabilization. Since radicals are surface states located within the gap of the semiconductor (Fig. 7a) only redox systems of a corresponding standard potential are suitable for a charge transfer via radicals. Others of much more positive potential, such as for instance $Ce^{3+/4+}$ (symbolized by donor II in Fig. 7a) do not stabilize[60]. This is of great disadvantage because it will be very difficult to produce stable photocells with high photovoltages. Certainly also other factors besides radical formation are of importance, such as surface morphology and film formation.

2.2.4 Influence of Surface Recombination and Trapping

Frequently it has been observed with n-type as well as with p-type electrodes in aqueous solutions that the onset potential of the pure photocurrent differs considerably from the flatband potential. The latter can be determined by capacity measurements in the dark as illustrated by the dashed line in the $i_{ph} - U_E$ curve in Fig. 8a. This effect is usually explained by recombination and trapping of minority carriers created by light excitation at the surface. It is obvious that these effects have a negative effect

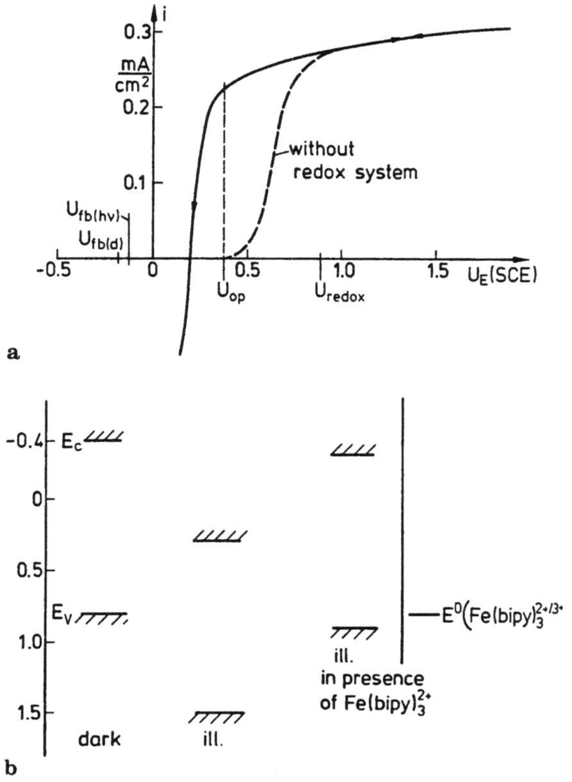

Fig. 8a and b. a) Current vs. potential for n-WSe_2 in 0.1 M H_2SO_4 with and without 0.1 M $Fe(bipy)_3SO_4$; **b)** Position of energy bands at the interface [66]

on the power output of a solar cell. The trapping or accumulation of minority carriers results in a shift of energy bands (Fig. 8b) at the interface as found with most semiconductor electrodes [64]. In the case of some layer compounds, such as e.g. WSe_2, the shift of bands is very large (~ 0.6 eV) and the flatband potential at illumination occurs very close to the onset potential of the photocurrent. Accordingly, the energy bands remain flat in the whole potential range between $U_{fb(dark)}$ and $U_{fb(h\nu)}$ [65]. It is assumed that the minority carriers are trapped in surface states located at steps. Since this effect does not occur in alkaline solutions it is reasonable to assume that insoluble WO_3 is formed at the steps which inhibits further transfer via these sites. Dissolution cannot really occur at the perfect areas of the smooth laminar surface because the W-atoms are screened by Se-atoms[49]; i.e. the surface is relatively inert.

The shift of energy bands upon illumination can nearly be avoided upon addition of a suitable redox system such as $[Fe(CN)_6]^{4-}$ [65]. Very recent investigations have shown that the same effect occurs in the presence of other redox systems having a relatively positive standard potential such as Fe^{3+} or $Fe(bipy)_3^{2+}$ (Fig. 8b)[66]. It is interesting to note that in all these cases the photocurrent onset occurs at much more negative potentials, i.e. rather close to $U_{fb(dark)}$. This result is important for a regenerative photocell because of two reasons: (i) In solar cells one chooses an operating point at which the power output is maximized. The corresponding operating voltage U_{op} will be in the potential range where a photocurrent occurs only in the presence of a redox system but not with the supporting electrolyte alone (Fig. 8a). Consequently the electrode should be stable at the operating potential even if the redox process cannot compete with the anodic decomposition at higher anodic potentials at which a photocurrent is also observed without a redox system. (ii) The fact that redox systems of a rather positive standard potential, such as e.g. $Fe(bipy)_3^{2+}$, can be used, should lead to rather high photovoltages.

According to these results it is favorable to apply semiconductor electrodes which exhibit a large overvoltage concerning the anodic decomposition but not for the oxidation of a redox couple. Unfortunately, there are only few examples which fulfil this condition. Another example was found with n-GaAs in the presence of Eu^{2+} [52]. In this case, however, the standard potential is very negative resulting in a very small photovoltage. In this respect single crystals of layer-compounds are favorable materials because of their special surface morphology.

2.3 Production of Storable Chemical Fuels

2.3.1 General Aspects

As already mentioned in the introduction the application of semiconductor/liquid systems is of special interest for the direct production of a chemical fuel by solar energy. Certainly the stability problem discussed in the previous sections is here even more severe. Therefore, on the first sight it may be easier to use a photovoltaic cell combined with an ordinary electrochemical cell in which a fuel is produced at inert metal electrodes. This argument could hold for the production of H_2 by electrolysis of H_2O although there are always energy losses in a two-cell system. It is not valid, however, in the case of the production of other fuels such as e.g. methanol and ammonia which have never been formed at metal electrodes. In principle semiconductor electrodes have the advantages that firstly charges can only be transferred via energy bands and secondly electrons in the conduction band may have a great reduction power and holes in the valence band a high oxidation power. It is then mainly a kinetic problem whether a certain fuel can be produced.

2.3.2 Principles of Photoelectrolysis of Water

The photoelectrolysis of H_2O can be performed in cells being very similar to those applied for the production of electricity. They differ only insofar as no additional redox couple is used in a photoelectrolysis cell. The energy scheme of corresponding systems, semiconductor/liquid/Pt, is illustrated in Fig. 9, the upper scheme for an n-type, the lower for a p-type electrode. In the case of an n-type electrode the hole created by light excitation must react with H_2O resulting in O_2-formation whereas at the counter electrode H_2 is produced. The electrolyte can be described by two redox potentials, $E^0(H_2O/H_2)$ and $E^0(H_2O/O_2)$ which differ by 1.23 eV. At equilibrium (left side of Fig. 9) the electrochemical potential (Fermi level) is constant in the whole system and it occurs in the electrolyte somewhere between the two standard energies $E^0(H_2O/H_2)$ and $E^0(H_2O/O_2)$. The exact position depends on the relative concentrations of H_2 and O_2. Illuminating the n-type electrode the electrons are driven toward the bulk of the semiconductor and reach the counter electrode via the external circuit at which they are consumed for H_2-evolution whereas the holes are directly

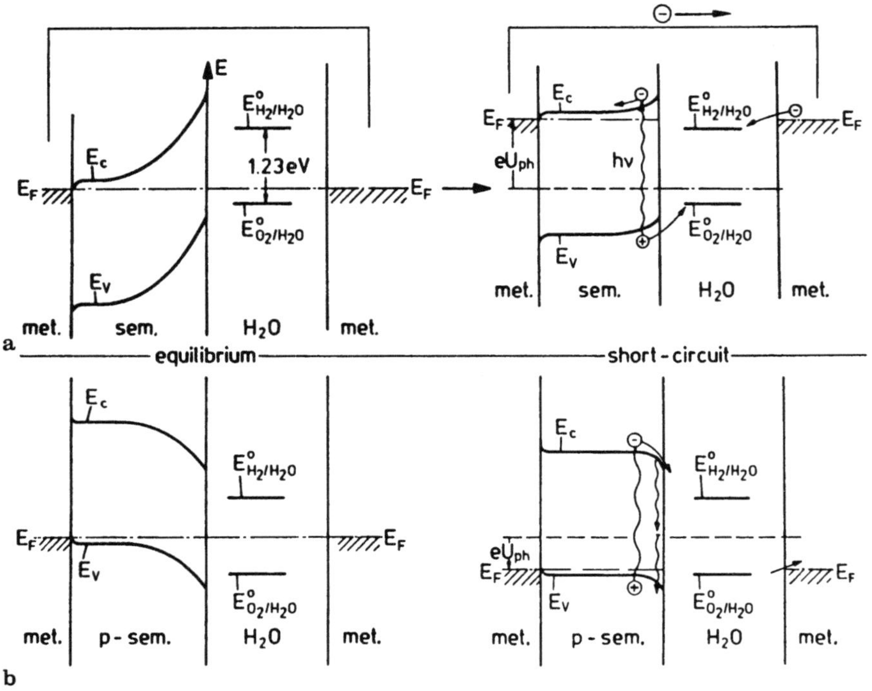

Fig. 9. Photocleavage of H_2O at n- and p-type electrodes (energy scheme)

transferred from the valence band to H_2O resulting into O_2-formation (right side of Fig. 9). These two processes can obviously only occur, if the bandgap is greater than 1.23 eV, the conduction band being negative of $E^0(H_2O/H_2)$ and the valence band positive of $E^0(H_2O/O_2)$. Since multielectron steps are involved in the reduction and oxidation of H_2O certain overvoltages occur for the individuel processes which leads to losses. Accordingly, the bandgap of the semiconductor must exceed sufficiently the minimum energy of 1.23 eV[2]. The first system being investigated was TiO_2[67]. The same conditions also hold for p-type electrodes. In this case — as shown in Fig. 9 — H_2 is formed at the semiconductor and O_2 at the counter electrode. Also other combinations have been suggested in which an n-type anode and p-type cathode are used[68,69] Such a system corresponds to a two photon system similar to that in plants.

As already mentioned before, the stability problem becomes very difficult in such a system because no redox system can be added. Consequently, only semiconductors can be used which do not undergo any corrosion. There are several oxide semiconductors of sufficient stability available. However, only few meet the energetic conditions mentioned above and all of them have a rather large bandgap. One example is $SrTiO_3$ (Eg = 3.3 eV) and in a corresponding photoelectrolysis cell H_2 and O_2 is formed, however, with a very low efficiency because only UV-light can be absorbed [76]. In the case of TiO_2 (rutile) the photocleavage of H_2O is also critical because the conduction band occurs only 0.1 eV above E^0 (H_2O/H_2) (see e.g. [7]). An interesting semiconducting material is n-RuS_2 (Eg = 1.3 eV) which is reported to be stable against corrosion although thermodynamically not expected[71]. Instead of anodic decomposition O_2-formation was observed under illumination. Tributsch explains this result by assuming that holes occur in d-states which do not represent bonding states, and that the strong coordination-type of bonding of intermediates formed in the H_2O-oxidation is less than the activation energy for photocorrosion[72]. Unfortunately, this material does not fulfil the energetic prerequisits derived above, i.e. the valence band occurs in the dark above $E^0(H_2O/O_2)$.

However, a large downward shift of energy bands by 1.8 eV[71] was found by illumination, similarly as that reported for WSe_2 (see Fig.8). The final position of the energy bands at the surface of RuS_2 is shown in Fig. 10. Interestingly, the photocurrent resulting in O_2 evolution occured at 1.2 V (SHE), the valence band being located, however, at +2.2 eV, i.e. considerably below the $E^0(H_2O/O_2)$-value. This implies that the oxidation of H_2O occurs via energy states within the gap, probably by energy states due to the formation of RuO_2 on the surface. Because of the large shift of bands upon illumination the conduction band occurs below $E^0(H_2O/H_2)$ so that no H_2O-cleavage is possible under short circuit conditions. Therefore, a corresponding cell can only be operated under an additional external bias as illustrated in Fig. 10.

In principle it would be interesting to use p-type semiconductors as photocathodes. Since then H_2 is produced at the p-semiconductor the stability should be no problem. Unfortunately, however, there are only few semiconductor materials which can be made to p-type electrodes and most of them do not fulfil the energetic requirements. In addition, strong surface recombination occurs which competes with the electron transfer[2,7]. Recently it has also been reported on photocleavage of H_2O by visible light at polycrystalline Fe_2O_3 p/n diode assemblies made of Mg- and Si-doped iron oxide.[73]. This result is rather surprising as the conduction band was normally found below $E^0(H_2O/H_2)$, i.e. H_2-formation should not be possible. The only way of explaining this result may be by the possibility that the energy bands of the heavily doped iron oxide have different positions.

During the last decade also many investigations with n-type electrodes were performed using sensitizers and catalysts. Corresponding results as obtained with catalysts will be discussed below.

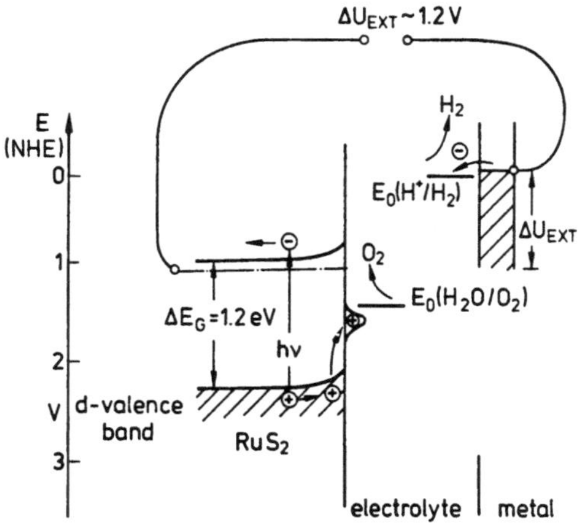

Fig. 10. Photocleavage of H_2O at n-RuS_2 under external bias[72]

2.3.3 Microheterogeneous Systems

Many investigations have been performed with semiconductor particles dissolved as colloids or used as suspensions in aqueous electrolytes. The principle advantage is the fact that a large semiconductor surface is available and that the photogenerated electrons and hole can reach the surface very quickly before they recombine. The question arises, however, which factors determine the reactivity of the charge carriers. Since usually a depletion layer exists at a semiconductor/liquid surface the energy bands are bent upward everywhere at the surface as illustrated in Fig. 11a. The thickness of such a space charge layer depends on the doping and on the potential across the space charge layer. It is given by[15]:

$$d_{sc} = \left(\frac{2\varepsilon\varepsilon_0 kT}{n_0 e^2}\right)^{1/2} \left(\frac{eU_{sc}}{kT} - 1\right)^{1/2} \quad (35)$$

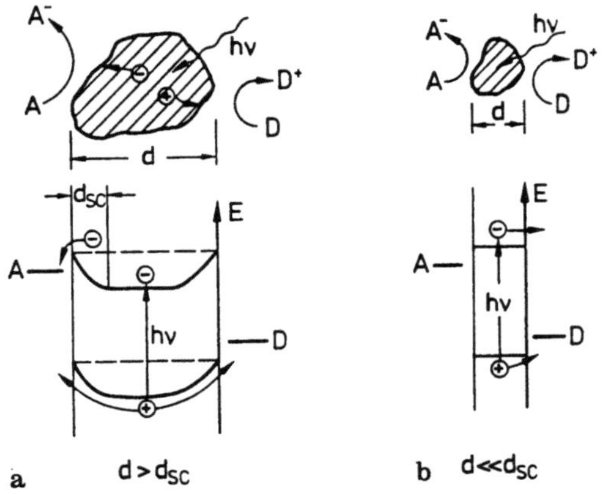

Fig. 11a and b. Electron and hole transfer at large **a)** and small **b)** semiconductor particels to an electron acceptor A and donor D

in which n_0 is the electron density in the bulk, ε the dielectric constant, ε_0 the permittivity in vacuum and U_{sc} the potential across the space charge layer. Taking typical values such as $\varepsilon = 10$; $n_0 = 10^{17}$ cm^{-3} and $U_{sc} = 1$ V the thickness is about 10^{-5} cm. In the case of much smaller particles of a diameter $d \ll d_{sc}$, of course, no space charge exists as shown in Fig. 11b.

Producing electron-hole pairs by light excitation in the small particles ($d < d_{sc}$) electrons and holes can easily be transferred to an electron acceptor and donor, respectively, provided that the energetic requirements are fulfilled. The quantum efficiency of the reaction depends on the transfer rate at the interface, on the recombination rate within the particle and on the transit time, the latter being given by[74]:

$$\tau = \frac{R^2}{\pi^2 D} \quad (36)$$

where R is the radius of the particle and D the diffusion coefficient of the excited charge carriers. Taking a typical value of $D \approx 0.1$ cm^2 s^{-1} and a radius R = 100 Å the average transit time will be only about 1 ps! This value is much smaller than that for recombination which is usually greater than 10 ns.

In the case of larger particles ($d > d_{sc}$) the sequence of reaction steps is more complex because of the space charge layer below the surface. Exciting a 1 μm-particle the minority carriers diffuse toward the surface and the transient time is then about 10 ns, i.e. recombination effects may be important. The holes reaching the surface are easily transferred because of the upward band bending (Fig. 11a). The electrons, however cannot leave the particle so that the latter will be charged up. This leads to a flattening of the bands so that finally also the electrons can cross the interface. Accordingly, the primary hole transfer leads to a compensation of the space charge. The latter can be approximated for a depletion layer by [15,79]:

$$Q_{sc} = (2\varepsilon\varepsilon_0 n_0 e)^{1/2}\, U_{sc}^{1/2} \tag{37}$$

Using $\varepsilon = 10$, $n_0 = 10^{17}\,\text{cm}^{-3}$ $U_{sc} = 1$ V and $R = 0.5\,\mu$ one obtains $Q_{sc} = 5 \times 10^{-15}$ As per particle and the number of charges forming the space charge layer are 3×10^4 per particle. Taking a suspension (10^8 particles of 1 μm in 1 cm^3) so that the incident light is just completely absorbed, then it takes in the average about 1 ns between the absorption of two photons in one particle for an incident photon flux of $10^{17}\,\text{cm}^{-2}\,\text{s}^{-1}$. Since at least 3×10^4 photons are required to compensate the space charge it should take around 30 ms before also the majority carriers can be transferred. Accordingly, there is no principle difference between small and big particles under stationary conditions. The situation changes, however, for extremely small colloidal particles because of changes in the electronic structure within the particles as e.g. intensively studied by Henglein and his group (see[76] and literature cited there).

On the other hand there is one further aspect which may be important for using large or small particles: As already mentioned above, in the average a time interval of 1 ns exists between the absorptions of two photons in a 1 μm-particles. Using a solution of particles with a size of 100 Å this time interval is about 100 μs, i.e. it is longer by a factor of 10^5 than that estimated for 1 μm-particles. This can be important insofar as in many reactions two or more electrons are involved as for example in the reduction of CO_2 and in the oxidation of alcohols or water. This means that an intermediate being formed in the first step (e.g. CO_2^- by reduction of CO_2) may undergo some other reaction before it captures another electron. From this point of view the use of larger particles is more favorable. On the other hand, a further electron required for the second reaction step may diffuse to another surface site at a 1 μm-particle because its surface is by a factor of 10^4 larger than that of a 100 Å-particle. Accordingly, it is difficult to predict an optimal particle size. Up to now there are no reports in the literature concerning this aspect of particle size.

It is interesting to note that some reactions proceed with large quantum yield, e.g. $\varphi = 0.8$ for the reduction of CO_2 to H_2CO_2 and the oxidation of SO_3^{2-} to SO_4^{2-} at ZnS-particles[71]. According to Henglein et al. high yields can be expected, if the reactions are ruled by different e^- and h^+ transfer mechanisms so that little interference between anodic and cathodic processes occurs [76].

2.3.4 Catalytic Processes

Many scientists have investigated various reactions at solid electrodes and particles on which catalysts were deposited in order to promote the reaction rate. Concerning

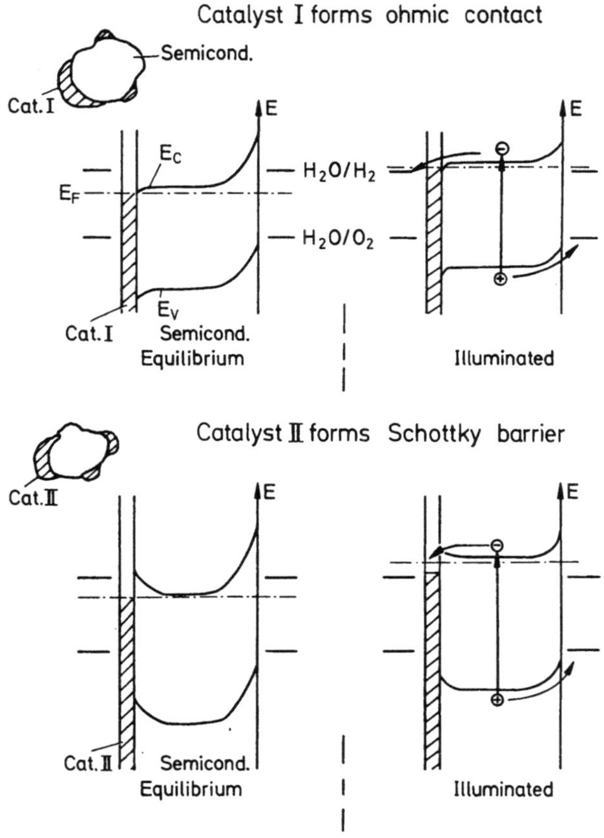

Fig. 12a and b. Reactions at catalyst loaded particles **a)** catalyst forming an ohmic contact **b)** forming a Schottky barrier

the nature of the catalytic effect it has to be realized at first that an additional semiconductor/catalyst interface is formed besides that of the semiconductor/liquid. Assuming that the metal deposited on a semiconductor particle forms an ohmic contact the system can be described by a simple energy scheme as given in Fig. 12a. In this microcell the Fermi level is constant everywhere in the system at equilibrium. Actually this system is in principle identical to a macrosystem in which two seperate electrodes, semiconductor and inert metal counter electrode, are used (compare with Fig. 9a). They differ only insofar as in the microcell the metal counter electrode is deposited directly onto the semiconductor. Applying this system to photoelectrolysis of H_2O, H_2 will be produced on the metal and O_2 on free sites of the n-type semiconductor when it is illuminated by light (right side of Fig. 12a).

In most cases, however, noble metals such as Pt, Ru and Rh or RuO_2 and Rh_2O_3 have been deposited on different semiconductors such as CdS[78-80], TiO_2[81-83], $SrTiO_3$[84] and WO_3[85] although these catalysts do not form an ohmic contact[86]. In connection with the photocleavage of H_2O mainly Grätzel introduced the idea of using semiconductor particles loaded even with two types of catalysts, one (e.g. Pt)

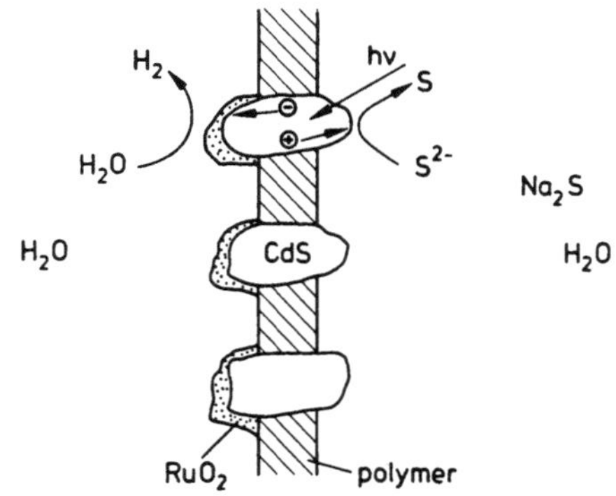

Fig. 13. Principle scheme of a semiconductor monograin membrane

for H_2-formation and another (e.g. RuO_2) for O_2-evolution[83]. The energy scheme for particles loaded with one catalyst forming a Schottky barrier is given in Fig. 12b. In this case one has two barriers, one at the catalyst-semiconductor and another at the semiconductor-liquid interface, assuming that the first one is lower than the other. Similarly, as for bare particles (compare with Fig. 11) the latter are negatively charged upon illumination because holes are transferred to the solution. This leads to recombination[87] and to a flattening of the energy bands so that finally electrons can also be transferred as illustrated in Fig. 12b. It should be emphasized here that the eletctrons must be transferred at that interface of the lowest barrier height, i.e. in this example across the catalyst. They would be transferred, however, across the free surface if the corresponding barrier is lower than that toward the catalyst.

This model was experimentally proved at first by Meissner et al. by using CdS monograin membranes[80] as illustrated in Fig. 13. In this technique single crystalline particles are fixed in a membrane in such a way that each particle can face the solution of both sides of the polymer membrane. This method makes it possible to investigate whether the cathodic and anodic processes occur at the catalyst or at the free surface. Corresponding measurments have shown that the H_2-evolution (majority carrier process) always occured at the catalyst independently of whether Pt or RuO_2 were used as a catalyst whereas the holes were transferred at the free surface either for anodic dissolution or the oxidation of an electron donor such as for instance S^{2-}-ions. In both cases barrier heights at the catalyst side are known, 1.6 eV for Pt[86] and 0.5 eV for RuO_2[88], which are both lower than that at the free surface. It may be surprising that H_2-evolution occurs at RuO_2 although it is an oxygen evolution catalyst. The result obtained with CdS monograin membranes shows clearly that reactions at catalyst loaded particles are primarily governed by the asymmetry of Schottky barriers formed on the semiconductor surfaces and not by any catalytic properties of the deposited material itself. This has been confirmed with RuO_2 loaded TiO_2-electrodes[89].

In this context it should be mentioned that the height of the Schottky barrier depends on the procedure of metal deposition and also on the pretreatment. Aspnes and Heller[86] have investigated for instance metal-semiconductor contacts produced by depositing Ru, Rh or Pt as 400 Å thick films. They found barrier heights for the metal in contact with air, of 0.6 eV for Ru on TiO_2, which decreased to zero in the presence of hydrogen. These results are consistent with those of Yamamoto et al.[90].

Similar results were obtained with $SrTiO_3$ but not with CdS[86]. In the case of p-InP the opposite effect was found, i.e. an increase of the barrier height upon admittance of H_2 to the metal. The nature of the ambient gas-induced changes were interpreted by a change in the surface dipole component of the metal work function[86]. The results obtained with CdS did not fit into this scheme probably because the surface chemistry of this material is rather complex[91].

As already mentioned above the photocleavage of H_2O has also been investigated at TiO_2- and CdS-particles loaded with two different catalysts such as Pt and RuO_2[78, 83]. In both cases it was originally reported that H_2- as well as O_2-evolution was observed. Especially the result that O_2 should be formed at a catalyst loaded CdS-particle found great interest because CdS shows normally anodic decomposition. Unfortunately nobody could really reproduce Grätzel's results. Frank originally repeated this experiment but he stated that his data were clearly insufficient to prove O_2-production[92]. Further investigations have shown that photocleavage of H_2O should be difficult or even impossible at CdS because of three reasons:

(i) If Pt and RuO_2 are deposited as catalysts on CdS one has three interfaces. According to values of the barrier heights at the RuO_2/CdS[80] and Pt/CdS[86] interfaces they differ at most by 1.1 eV as illustrated in Fig. 14, so that H_2O splitting is not possible. (ii) At the bare CdS-surface the position of the valence band occurs already above $E_0(H_2O/O_2)$[62]. (iii) In the presence of O_2 in the solution the anodic decomposition of CdS is strongly enhanced, a process which can be quantitatively described by the overall reaction[93, 94]

$$CdS + 2\,O_2 \xrightarrow{\text{light}} Cd^{2+} + SO_4^{2-} . \tag{38}$$

Already Henglein found with bare CdS colloids that no reaction occurs in O_2-free solutions probably because a blocking sulfur layer was formed[95]. This was also confirmed with catalyst loaded particles[62].

In connection with this problem it should be mentioned that O_2-formation was found at CdS electrodes coated with polypyrrole and RuO_2[96] under anodic polarization whereby the anodic decomposition could be considerably reduced. Under open circuit conditions only H_2-evolution was observed, whereas O_2 could obviously not be detected. This result is not in contradiction to the first experiment because the Fermi level can pass the electrochemical potential of H_2O/O_2 under bias. Very recently it was reported on photocleavage of H_2O at catalyst loaded CdS-particels in the

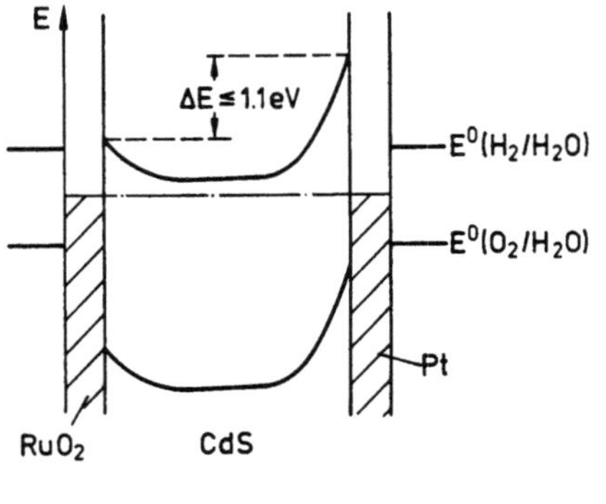

Fig. 14. Energy scheme of semiconductor particle loaded with two different catalysts

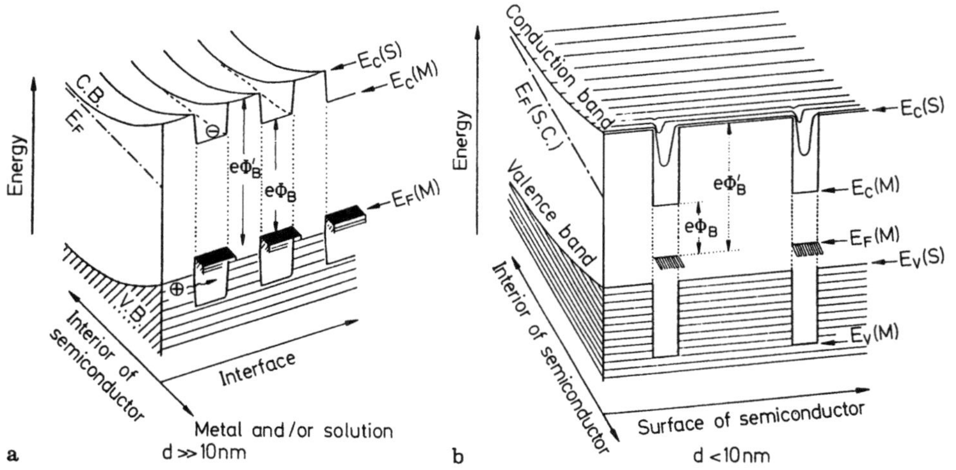

Fig. 15. Schematic energy diagram of the n-type semiconductor electronic bands at the solid/liquid interface modulated by discontinuous metal coating[99]

presence of $[\{Ru(OH)(edta)\}_2(O_2)]$[97]. In this paper it was also mentioned that small traces of O_2 were found without the peroxycomplex, a result which is certainly due to some small air leak.

Concerning the function of catalysts the question arises what happens when metal clusters deposited on semiconductor electrodes become smaller and smaller. The resulting effects have been analyzed by Nakato and Tsubomara[98,99]. These authors have developed a corresponding model as illustrated in Fig. 15. In the case of an n-type electrode the energy bands are modulated as shown in Fig. 15a if metal clusters of a diameter of at least 10 nm are deposited. The barrier height $e\varphi_B$ underneath the metal is entirely determined by the semiconductor/metal contact. It is independent of the electrolyte used in the experiment. The barrier height of the free surface, however, depends on the redox couple used in the solution. If the width of the free surface area is not too large, photoholes reaching the surface can enter the metal via quantum mechanical penetration through the potential well as indicated by an arrow in Fig. 15a. The maximum photovoltage is determined by $e\varphi_B$ and the photocurrent is only carried by electrons transferred via the metal. In the case of much smaller islands, i.e. $\leqq 5$ nm, the quantitized levels of the electrons in the potential well in the conduction band should become considerably higher than the bottom of the well (Fig. 15b) so that the electrons have a higher kinetic energy. Since the potential well is so narrow the band modulation diminished rapidly toward the interior of the crystal. These two effects make the effective barrier height little different of that of the naked surface (Fig. 15b), i.e. the barrier height $e\varphi'_B$ determines the photovoltage. Since $e\varphi'_B$ depends on the standard potential potential U^0_{redox} of the redox couple the photovoltage should depend on U^0_{redox}. This was actually found with n-Si electrodes[99,33] and very high photovoltages of up to 0.68 V at $i_{ph} = 20$ mA/cm^2 were obtained in a corresponding photovoltaic cell (see also Table 1).

This is a very important result which should also be of significance in the preparation of catalysts on semiconductor surfaces for the production of a chemical fuel. The question arises, however, whether such small metal clusters ($\leqq$ 5 nm) still exhibit catalytic properties and how many metal atoms form a cluster of metallic properties. Investigations with colloidal silver gave a result of 12 atoms/particles[76)].

All these results indicate that one is just at the beginning of understanding the function of catalysts being deposited on a semiconductor. There is still quite a confusion in many papers published in this field. Therefore the catalytic properties depend so much on the procedure of deposition[100)]. It seems to be rather difficult to produce a catalyst for O_2-formation, as shown by results obtained with TiO_2 (see e.g.)[101)]. Rather recently new concepts for the synthesis of new catalysts have been developed applicable for multielectron transfer reactions. Examples are transition metal cluster compounds such as $Mo_{4.2}Ru_{1.8}Se_8$[102)] and di- and trinuclear Ru-complexes[103)].

2.3.5 Photoelectrolysis of H_2S and HI

As lined out in the previous section, great difficulties occur in the photocleavage of H_2O by visible light. In the case of H_2S-photocleavage the situation is much simpler because the oxidation of S^{2-}-ions can compete sufficiently fast with the anodic decomposition. This process was at first verified by Nozik[68b)] using CdS/Pt photodiodes. Further investigations were performed with CdS-suspensions, the particles being loaded with RuO_2[104)] or Pt[105)] as catalysts. The effect of RuO_2 was originally attributed to the catalysis of hole transfer from the valence band of CdS to the S^{2-}-ions in the solution[104)]. Lateron it was shown by using CdS monograin membranes (Fig. 13) that not the hole transfer but the electron transfer is catalyzed by RuO_2 because the H_2 evolution was found at this catalyst as shown in Fig. 16[80)]. Accordingly

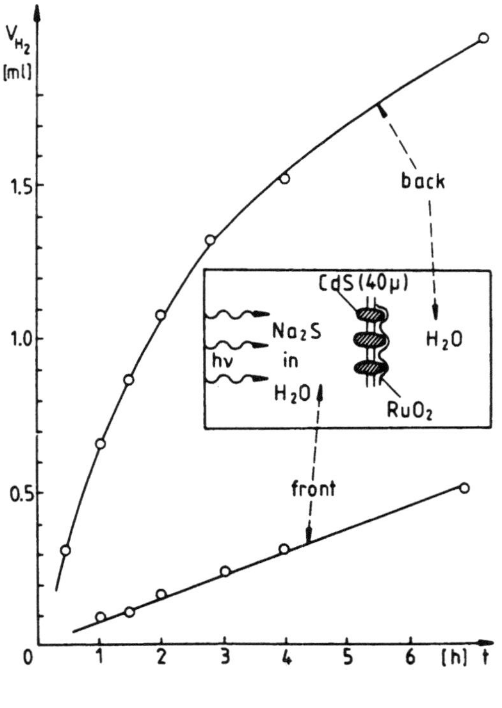

Fig. 16. Volumn of H_2 produced by photocleavage of H_2S at a CdS monograin membrane vs. irrdiation time[80)]

we have

$$2\,H_2O + 2e^- \rightarrow H_2 + 2\,OH^- \qquad \text{at Pt or } RuO_2 \text{ and} \qquad (39)$$
$$S^{2-} + 2h^+ \rightarrow S^0 \qquad \text{at the free surface}. \qquad (40)$$

Quantum efficiencies of up to 30% have been obtained.

This result was interpreted by the formation of a Schottky barrier at the CdS/RuO_2-interface as already discussed in the previous section. The H_2-production at CdS/RuO_2-suspensions could be considerably increased by addition of sulfite because the latter served as a sink for sulfur produced via reaction (40) [106].

During the last couple of years CdS-containing Nafion membranes have been applied for the photocleavage of H_2S[107]. They are not comparable with the monograin membranes because the CdS particles are at randomly distributed in a rather thick Nafion membrane. This technique is attractive for some applications because the semiconductor particles are immobilized[108]. On the other hand, problems may arise because of diffusion problems in the nafion membrane. Mainly the photoassisted H_2-formation at CdS was investigated in the presence of a Pt-catalyst and with co-precipitated ZnS · CdS without a catalyst[109].

The photoelectrolysis of HI was studied only with cells containing extended electrodes such as p-InP and an inert metal counter electrode[110]. In this case a Schottky barrier was formed on the InP-surface by deposition of Rh which is important for avoiding a large overvoltage concerning the onset potential of the photocurrent (Fig. 17). Under short circuit conditions the anodic and cathodic currents can reach considerable values during light excitation. This may be an interesting system, if it is combined with a fuel cell because the fuels, H_2 and I_3^-, can be stored.

2.3.6 Reduction of CO_2 and Formation of CH_3OH

Various attempts have been made to produce methanol by photoreduction of CO_2 at semiconductor electrodes and particles (see e.g.)[111]. In principle there is a good chance to produce methanol at a semiconductor because the conduction band can be

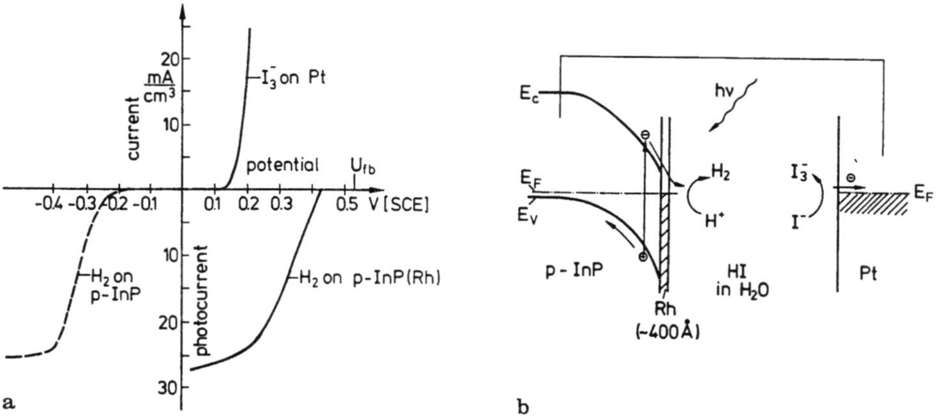

a b

Fig. 17a and b. Photoelectrolysis of HJ at p-InP loaded with Rh. **a)** partial currents[110]; **b)** energy diagram sufficiently negative[112]. With metal electrodes CO_2 can also be reduced but only formic acid was formed. The same product was found with particles during illumination using sulfite as a hole scavanger[113].

Very interesting are some investigations performed with GaAs-electrodes by Frese and Canfield[114]. They have observed that CO_2 is selectively reduced to CH_3OH at the (111)As-face during cathodic polarization. Current efficiencies of 100% were obtained. It is important to note that this result was only found by using reagent grade salts and simply distilled H_2O. In the case of much purer chemicals (99.999% Na_2SO_4 $1.7 \times 10^7\,\Omega$cm H_2O) practically no CH_3OH was obtained. This surprising result may be due to the deposition of traces of impurities (Zn, Cu, As and Ru) which catalyze somehow the CH_3OH-formation[115]. The authors proposed a reaction mechanism as follows:

$$nH^+ \times ne^- \rightarrow nH_{ad} \quad (41a)$$
$$2\,H_{ad} \rightarrow H_2 \quad (41b)$$
$$CO_2 + e^- + H^+ \rightarrow \cdot COOH_{ad} \quad (41c)$$
$$H_{ad} + \cdot COOH_{ad} \rightarrow I \rightarrow CH_3OH \quad (41d)$$

in which I represents some unknown intermediate. Other authors[116] have mentioned that CH_3OH is also formed by illuminating p-GaAs under open circuit conditions.

As already mentioned before the electrochemical reduction of CO_2 at a metal electrode leads only to the formation of formic acid. Recently it has been reported by Ogura et al. (see [117] and literature cited therein), however, that at a Pt-electrode coated by a layer of Everitt's salt (ES), $K_2Fe(II)[Fe(II)(CN_6)]$, CO_2 is selectively reduced to methanol in the presence of metal complexes as homogeneous catalysts and a primary alcohol. The overall reaction is given by

$$CO_2 + 6\,ES + 6\,H^+ \rightarrow CH_3OH + 6\,PB + 6\,K^+ + H_2O \quad (42)$$
$$5\,PB + 6\,e^- \rightarrow 6\,ES \quad (43)$$

in which PB (Prussian Blue) is the oxidized species of ES. A typical catalyst was $Na_3[Fe(CN)_5(H_2O)]$ in which a coordination bond is formed between the labile ligand and the primary alcohol. Ogura et al. have clearly shown that methanol was produced when ethanol was used as a primary alcohol. Surprisingly, they also reported that considerably higher current efficiencies have been obtained by using methanol itself as a primary alcohol. In the latter case, however, the increase of the methanol concentration upon cathodic polarization was still rather small (only 3% within 3 hours) although current efficiencies of up to 100% are reported. Unfortunately, the authors did not give any information on the accuracy of measuring small changes of the methanol concentration. The same research group has also combined the modified Pt-electrode with a semiconductor photoanode (TiO_2) and found methanol formation during illumination[118]. These are interesting results; however, more investigations are required before it can be judged whether this method is really suitable for the production of methanol from CO_2. It should be further mentioned that also other mediators, such as Co-complexes, have been applied for the reduction of CO_2[119].

2.3.7 Photoelectrochemical production of NH_3

The photolytic reduction of N_2 at TiO_2-suspensions was at first reported by Schrauzer et al.[12,121]. Small amounts of NH_3 and N_2H_4 were obtained as products. The highest activity was found with anatase containing 20–30% rutile. Very low yields were also obtained with p-GaP electrodes under illumination[122]. It is much easier to produce NH_3 from NO_2^--solutions at CdS- and TiO_2-particles using S^{2-}-ions as hole scavengers[123]. Efficiencies are not reported yet. Recently the formation of NH_3 from NO was observed at p-GaAs electrodes under illumination[124]. In this case NH_3-formation was only found in the presence of transition metal ions or their complex with EDTA.

This research is still in a rather premature state because of the problem of nitrogen fixation.

3 Conclusion

As already mentioned in the introduction, various fundamental and many empirical results have been published. Although photoelectrochemical cells are easily made, many problems concerning the stability of semiconductors and the function of catalysts still remain to be solved. There are few other approaches such as for instance sensitization (see e.g.)[125] which are not treated here. In addition it should be mentioned that photoelectrochemical systems have been used for light induced synthesis of organic compounds (see e.g.)[126,127] which could also not be considered in this article.

4 References

1. Harris, L. A., Wilson, R. H.: Ann. Rev. Mat. Sci. B, 99 (1978)
2. Gerischer, H.: in 'Solar Energy Conversion', Topics in Appl. Phys. (Seraphin, B. O., ed.) Vol. 31, p. 115, Berlin Springer Verlag 1979
3. Memming, R.: Phil. Techn. Rev. *38*, 160 (1978/79)
4. Nozik, A. J.: Ann. Rev. Phys. Chem. *29*, 189 (1978)
5. Rajeshwar, K., Singh, P., Dubow, J.: Electrochim. Acta *23*, 1117 (1978)
6. Butler, M. A., Ginley, D. S.: J. Mat. Sci. *15*, 1 9180)
7. Memming, R.: Electrochim. Acta *25*, 77 (1980)
8. Heller, A.: Accts. Chem. Res. *14*, 154 (1981)
9. Tribusch, H.: in Sol. Energy Mat., Structure and Bonding Vol. 49, p. 127, Berlin Springer Verlag 1982
10. Bard, A. J.: Science *207*, 139 (1980)
11. Grätzel, M.: in [9], p. 37
12. Lewis, N. S.: Ann. Rev. Mat. Sci. *14*, 95 (1984)
13. Pleskow, YU. V.: Progr. in Surf. Sci. *15*, 401 (1984)
14. Gerischer, H.: in Physical Chemistry (Egring, M., Jost, W., Henderson, D., eds.) Vol. IV A, p. 463, New York Academic Press 1970
15. Memming, R.: in Comprehensive Treatise of Electrochemistry (Conway, B. E., Bockris, J. O'M., Yeager, E., Eds.) Vol. 7, p. 529, New York, Plenum Press 1983
16. Pleskow, YU, V.: in [15] Vol. 1, p. 291, 1980
17. Morrison, S. R.: Electrochemistry at Semiconductor and Oxidized Metal Electrodes, New York, Plenum Press 1980
18. Gomes, W. P., Cardon, F.: Progr. in Surf. Sci. *12*, 155 (1982)
19. Buhl, M. L., Bird, R. E., Bilchak, R., Connolly, J. S., Bolton, J. R.: Sol. Energy (USA) *32*, 75 (1984)

20. Fan, J. C. C., Bosler, C. O., Chapman, R. L.: Appl. Phys Lett. *32*, 390 (1978)
21. Sze, S. M.: Physics of Semiconductor Devices 2nd. Ed., New York, John Wiley 1981
22. Reineke, R., Memming, R.: Surf. Sci. submitted
23. Memming, R.: Ber. Bunsenges. Phys. Chem. *91*, 353 (1987)
24. Heller, A., Schwartz, G. P., Vadimsky, R. G., Menezes, S., Miller, B.: J. Electrochem. Soc. *125*, 1156 (1978)
25. Licht, S., Tenne, R., Dagan, G., Hodes, G., Manassen, J., Triboulet, R., Rioux, J., Levy-Clement, C.: Appl. Phys. Lett *46*, 608 (1985)
26. Chang, K. C., Heller, A., Schwartz, B., Menezes, A., Miller, B.: Science *196*, 1097 (1977)
27. Parkinson, B. A., Heller, A., Miller, B.: J. Electrochem. Soc. *126*, 954 (1979)
28. Nakatani, K., Matsudaira, S., Tsubomura, H.: ibid. *125*, 406 (1978)
29. Cahen, D., Chen, Y. W.: Appl. Phys. Lett. *45*, 746 (1984)
30. Lewerenz, H. J., Goslowsky, H., Husemann, K. D., Fiechter, S.: Nature *321*, 687 (1986)
31. Gobrecht, J., Gerischer, H., Tributsch, H.: J. Electrochem. Soc. *125*, 2085 (1978)
32. Tenne, R., Wold, A.: Appl. Phys. Lett. *47*, 707 (1985)
33. Nakato, Y., Yano, H., Tsubomura, H.: Chemistry Lett., p. 987 (1986)
34. Rosenbluth, M., Lewis, N. S.: J. Am. Chem. Soc. *108*, 4689 (1986)
35. Casagrande, L. G., Lewis, N. S., Kelso, S. M.: Electrochem. Soc. Meeting, San Diego, Abstract No. 607 (1986)
36. Heller, A., Miller, B., Thiel, F. A.: Appl. Phys. Lett. *38*, 282 (1981)
37. Hodes, G.: J. Photochem. *29*, 243 (1985); in Energy Resources Through Photochemistry and Catalysis (Grätzel, M., ed.) p. 521, New York, Academic Press 1983
38. Licht, S., Manassen, J.: J. Electrochem. Soc. *132*, 1076 (1985)
39. Licht, S., Tenne, R., Flaisher, H., Manassen, J.: ibid. *133*, 52 (1986)
40. Menezes, S., Lewerenz, H. J., Bachmann, K. J.: Nature *305*, 615 (1983)
41. Lewerenz, H. J., Kötz, E. R.: J. Appl. Phys. *60*, 1430 (1986)
42. Ennaoui, A., Tributsch, H.: J. Electroanal. Chem. *204*, 185 (1986)
43. Ennaoui, A., Fiechter, S., Jaegermann, W., Tributsch, H.: J. Electrochem. Soc. *133*, 97 (1986)
44. Hodes, G., Manassen, J., Cahen, D.: ibid. *127*, 544 (1980)
45. Nelson, R. J., Williams, J. S., Leamy, H. J., Miller, B., Casey, H. C., Parkinson, B. A., Heller, A.: Appl. Phys. Lett. *36*, 76 (1980)
46. Tubbesing, K., Meissner, D., Memming, R., Kastening, B.: J. Electroanal. Chem. *214*, 685 (1986)
47. Kautek, W., Gerischer, H.: Electrochim. Acta *27*, 355 (1982)
48. Gobrecht, J., Tributsch, H., Gerischer, H.: J. Electrochem. Soc. *125*, 2085 (1978)
49. Lewerenz, H. J., Gerischer, H., Lübke, M.: ibid. *131*, 100 (1984)
50. Lewerenz, H. J., Heller, A., DiSalvo, F. J.: J. Am. Chem. Soc. *102*, 1877 (1980)
51. Kelly, J. J., Memming, R.: J. Electrochem. Soc. *129*, 730 (1982)
52. Meissner, D., Sinn, Ch., Memming, R., Notten, P. H. L., Kelly, J. J.: in Homogeneous and Heterogeneous Photocatatlyses (E. Pelizzetti, Serpone, A. N., eds.) p. 317, Dordrecht, D. Reidel Publ. Comp. 1986
53. Memming, R.: J. Electrochem. Soc. *125*, 117 (1978)
54. Frese, K. W., Madou, M. J., Morrison, S. R.: J. Phys. Chem. *84*, 3172 (1980); J. Electrochem. Soc. *128*, 1528 (1981)
55. Overmeire, F. v., Kerchove, F. v. d., Gomes, W. P., Cardon, F.: Bull. Chem. Soc. Belg. *89*, 181 (1980)
56. Gomes, W. P., Overmeire, F. v., Vanmaekelbergk, D., Kerchove, F. v. d., Cardon, F.: in Photoeffects at Semiconductor-Electrolyte Interfaces, ACS-Series 145 (Nozik, A. J., ed.) 120 (1981)
57. Menezes, S., Miller, B.: J. Electrochem. Soc. *130*, 517 (1983)
58. Gerischer, H., Lübke, M.: Ber. Bunsenges. Phys. Chem. *87*, 123 (1983)
59. Cardon, F., Gomes, W. P., Kerchove, F. v. d., Vanmaeckelbergk, D., Overmeire, F. v.: Far. Discussions *70*, 153 (1980)
60. Memming, R.: in Photoelectrochemistry, Photocatalysis and Photoreactors (Schiavello, M., ed.) p. 107, Dordrecht, D. Reidel Publ. Comp. 1985
61. Benito, R. M., Nozik, A. Y.: J. Phys. Chem. *89*, 3429 (1985)
62. Meissner, D., Memming, R., Kastening, B., Bahnemann, D.: Chem. Phys. Lett. *127*, 419 (1986)

63. Kelly, J. J., Ven, J. v. d., Meerakker, J. E. A. M. v. d.: J. Electrochem. Soc. *132*, 3026 (1985)
64. Memming, R., Kelly, J. J.: in Photochemical Conversion and Storage of Solar Energy (Connally, J. S., ed.) p. 243, New York Academic Press 1981
65. McEvoy, A. J., Etman, M., Memming, R.: J. Electroanal. Chem. *190*, 225 (1985)
66. Sinn, Ch., Memming, R.: to be published
67. Fujishima, A., Honda, K.: Nature *238*, 37 (1972)
68. Nozik, A. J.: (a) Appl. Phys. Lett. *29*, 150 (1976); (b) Appl. Phys. Lett. *30*, 567 (1977)
69. Fornarini, L., Nozik, A. J., Parkinson, B. A.: J. Phys. Chem. *88*, 3238 (1984)
70. Mavroides, J. G., Kafalos, J. A., Kolesar, D. F.: Appl. Phys. Lett. *28*, 241 (1976)
71. Kühne, H. M., Tributsch, H.: J. Electroanal. chem. *201*, 263 (1986)
72. Tributsch, H.: J. Photochem. *29*, 89 (1985)
73. Leygraf, Ch., Hendewerk, M., Somorjai, G. A.: Proc. Nat. Acad. Sci. *79*, 5739 (1982); J. Phys. Chem. *86*, 4484 (1982)
74. Grätzel, M., Frank, A. J.: J. Phys. Chem. *86*, 2964 (1982)
75. Bard, A. J., Bocarsky, A. B., Fan, F. R., Walton, E. G., Wrighton, M. S.: J. Am. Chem. Soc. *102*, 3671 (1980)
76. Henglein, A.: in Modern Trends of Colloidal Science in Chemistry and Biology' (Eicke, H. F., ed.) p. 126, Stuttgart, Birkhäuser Verlag 1985; Topics in Current Chemistry, in press
77. Henglein, A., Gutiérrez, M., Fischer, M.: Ber. Bunsenges. Phys. Chem. *88*, 170 (1984)
78. Kalyanassundaram, K., Borgarello, E., Grätzel, M.: Helv. Chim. Acta *64*, 362 (1981)
79. Darwent, J. R., Porter, G.: J. Chem. Soc. Chem. Comm. 145 (1981)
80. Meissner, D., Memming, R., Kastening, B.: Chem. Phys. Lett. *96*, 34 (1983)
81. Sato, S., White, J. M.: ibid. *72*, 83 (1982)
82. Kawai, T., Sakata, T.: ibid. *72*, 87 (1982)
83. Duonghong, D., Borgarelo, E., Grätzel, M.: J. Am. Chem. Soc. *103*, 6324 (1981)
84. Lehn, J. M., Savage, J. P., Ziessel, R.: Nouv. J. Chim. *4*, 623 (1980); *5*, 291 (1981)
85. Erbs, W., Desilvestro, J., Borgarello, E., Grätzel, M.: J. Phys. Chem. *88*, 4001 (1984)
86. Aspnes, D. E., Heller, A.: J. Vac. Sci. Technol. *B1*, 602 (1983); J. Phys. Chem. *87*, 4919 (1983)
87. Gerischer, H.: J. Phys. Chem. *88*, 6096 (1984)
88. Gissler, W., McEvoy, A. J.: J. Appl. Phys. *53*, 1251 (1982)
89. Sakata, T., Hashimoto, K., Kawai, T.: J. Phys. Chem. *88*, 5214 (1984)
90. Yamamoto, N., Tonomura, S., Matsuoka, T., Tsubomura, H.: Surf. Sci. *92*, 400 (1980); J. Appl. Phys. *52*, 6230 (1981)
91. Meissner, D., Benndorf, C., Memming, R.: Appl. Surf. Sci. *27*, 423 (1987)
92. Frank, A. J.: private communication
93. Meissner, D., Memming, R., Shuben, Li, Yesodharan, S., Grätzel, M.: Ber. Bunsenges. Phys. Chem. *89*, 121 (1985)
94. Meissner, D., Memming, R., Kastening, B.: J. Phys. Chem., subm.
95. Henglein, A.: Ber. Bunsenges. Phys. Chem. *86*, 301 (1982)
96. Frank, A. J., Honda, K.: J. Phys. Chem. *86*, 1933 (1982); in Energy Resources through Photochemistry and Catalysis (Grätzel, M., ed.), New York, Academic Press 1983
97. Taqui Khan, M. M., Bhardway, R. C., Jadhav, C. M.: J. Chem. Soc., Chem. Comm. 1690 (1985)
98. Nakato, Y., Ueda, K., Tsubomura, H.: J. Phys. Chem., in press
99. Tsubomura, H., Nakato, Y.: Nouv. J. Chim. *11*, 167 (1987)
100. Uosaki, K., Yoneda, R., Kita, H.: J. Phys. Chem. *89*, 4042 (1985)
101. Kiwi, J., Grätzel, M.: ibid. *88*, 1302 (1984)
102. Vante, N. A., Tributsch, H.: Nature *323*, 431 (1986)
103. Ramaraj, R., Kira, A., Kaneko, M.: Ang. Chemie *98*, 824 (1986); *98m*, 1012 (1986)
104. Borgarello, E., Kalyanasundaram, K., Grätzel, M.: Helv. Chim. Acta *65*, 243 (9182)
105. Matsumura, M., Saho, Y., Tsubomura, H.: J. Phys. Chem. *87*, 3807 (1983)
106. Thewissen, D. H. M. W., Tinnemans, A. H. A., Eenwhorst. Reinten, M., Timmer, K., Mackor, A.: Nouv. J. Chim. *7*, 191 (1983)
107. Mau, A. W., Huang, Ch., Kakuta, N., Bard, A. Y., Campion, A., Fox, M. A., White, Webber, S. E.: J. Am. Chem. Soc. *106*, 6537 (1984)
108. Kakuta, N., White, J. M., Campion, A., Bard, A. J., Fox, M. A., Webber, S. E.: J. Phys. Chem. *89*, 48 (1985)

109. Kakuta, N., Park, K. H., Finlayson, M. F., Ueno, A., Bard, A. Y., Campion, A., Fox, M. A., Webber, S. E., White, J. M.: ibid. *89*, 732 (1985)
110. Levy-Clement, C., Heller, A., Bonner, W. A., Parkinson, B. A.: J. Electrochem. Soc. *129*, 1701 (1982)
111. Halman, M.: in Energy Resources through Photochemistry and Catalysis (Grätzel, M., ed.) p. 507, New York, Academic Press 1983
112. Honda, K., Fujishima, A., Watanabe, T.: in Solar-Hydrogen Energy Systems (Ohta, T., ed.) p. 137, Oxford, Pergamon Press 1981
113. Henglein, A.: Pure Appl. Chem. *56*, 1215 (1984)
114. Frese, K. W., Canfield, D.: J. Electrochem. Soc. *130*, 1772 (1983); *131*, 2518 (1984)
115. Frese, K. W.: private communication
116. Sears, W. M., Morrison, S. R.: J. Phys. Chem. *89*, 3295 (1985)
117. Ogura, K., Takamagari, K.: J. Chem. Soc. Dalton Trans, 1519 (1986)
118. Ogura, K., Takagi, M.: Sol. Energy *37*, 41 (1986)
119. Tinnemans, A. H. A., Kosters, Thewissen, D. H. M. W., Mackor, A.: Recl. Trav. Chim. *103*, 288 (1984)
120. Schrauzer, G. N., Guth, T. D.: J. Am. Chem. Soc. *99*, 7189 (1977)
121. Schrauzer, G. N., Guth, T. D., Salehi, J., Strampach, N., Liu Nan Hui, Palmer, M. R.: in [52], p. 509
122. Grayer, S., Halman, M.: J. Electroanal. Chem. *170*, 363 (1984)
123. Halmann, M., Zuckerman, K.: in [52], p. 52L
124. Ogura, K., Takagi, M.: J. Electroanal. Chem. *183*, 277 (1985)
125. Memming, R.: Progr. Surf. Sci. *17*, 7 (1984)
126. Sakata, T., Kawai, T.: in [111], p. 331
127. Fox, M. A.: in [52], p. 363; Topics in Current Chemistry, in press

Mechanism of Reactions on Colloidal Microelectrodes and Size Quantization Effects

Arnim Henglein

Hahn-Meitner-Institut Berlin Bereich Strahlenchemie, D-1000 Berlin 39, FRG

Table of Contents

Small particles of metals in solution often behave like electrodes although they are not connected to a battery which determines their potential. However, when a chemical reaction occurs in the solution of such particles intermediate free radicals may transfer electrons to them. The particles are thus charged chemically and are able to act as a metal electrode on cathodic potential. Electron transfer reactions become possible at these micro-electrodes which cannot be brought about by the radicals in the absence of the colloidal catalyst.

Small semiconductor particles also act like microelectrodes upon illumination. Electrons and positive holes are created in the particles which initiate redox reactions. The charge carriers may also recombine and emit fluorescence light. Reaction with a solute leads to quenching of the fluorescence.

Besides these chemical effects, which are understood in terms of the established theories in semiconductor physics and chemical kinetics, new physico-chemical phenomena are observed in the case of extremely small particles. The metal or semiconductor behavior is gradually lost with decreasing size, the consequences being drastic changes in the optical properties of the materials and also in their photocatalytic effects.

1 Introduction

Conventional colloid chemistry and electrochemistry have always been closely related with each other, the keywords electrophoresis, double layer theory, and specific adsorption describing typical aspects of this relationship. In more recent times, new aspects have arisen which again bring colloid chemistry into contact with modern developments in electrochemistry, particularly in photoelectrochemistry. These aspects include the use of colloidal particles as catalysts for electron transfer reactions and as photocatalysts. In fact, the similarity between the reactions that occur on colloidal particles and on compact electrodes has often been emphasized by calling the small particles "microelectrodes".

In the course of this development a trend for the preparation of smaller and smaller colloidal particles both in solution and in the solid state has become evident. There are two reasons for this trend. Firstly, small particles attenuate light almost exclusively by absorption and not by scattering, i.e. the powerful optical methods can be applied which are generally used in kinetic studies, especially for the detection of short-lived intermediates. Secondly, in catalysis in general, smaller particles or clusters of molecules are being thoroughly investigated today both in the solid and the gaseous phase [1]. The research on small colloidal particles in solution complements these studies. The electronic properties of metals and semiconductors are not properties of single atoms or molecules, but are brought about by the periodic array of a large number of atoms or molecules in a crystal lattice. A transition from metal or semiconductor properties to molecular properties is observed as smaller and smaller particles are investigated. In this transition range the optical, electrochemical and catalytic properties of the materials may change drastically.

Conventional colloid chemistry was developed during the first decades of this century. When Wo. Ostwald [2] wrote his famous book "Die Welt der vernachlässigten Dimensionen" (The world of the neglected dimensions) he wanted to draw the attention of his fellow chemists to this new field of chemistry. In this book, he also defined an approximate size range of colloidal particles, but immediately went on to say that there should be a gradual transition from colloids to molecules. This transition became forgotten in the further development of colloid chemistry, although colloid chemists use "seeds" for nucleation to produce larger particles. It seems that a systematic investigation of "seeds", which fall into the transition range between semiconductor and molecular properties, has never been carried out. If Ostwald were to write a book on modern problems in colloid chemistry today, he would perhaps choose the attractive old title, but point to the 1 nm size as the modern neglected dimension. Work is now being done on such small particles with increasing intensity. Our article is being written at a time of rapid developments, in part characterized by provisory experiments and theories. We do not attempt to present here a complete review but try to focus on a few topics that seem to be typical for catalysis and photocatalysis by colloids and size quantization effects in colloidal particles.

In studies of this kind, methods developed in radiation chemistry and photochemistry are often applied [3,4]. The methods of pulse radiolysis [5,6] and flash photolysis [7] allow one to investigate the mechanism of reactions in which free radicals, electrons and positive holes are the intermediates. In order to understand the mechanisms of processes that occur on colloidal particles it is important to know how free radicals interact with these particles. In pulse radiolysis studies, the radicals are produced in the bulk solution by a short pulse of high energy radiation. The reactions of the radicals with the colloid are observed after the pulse by recording the accompanying changes in optical absorption or in conductivity. The concentration in which the radicals are formed does not generally exceed 10^{-6} M thereby avoiding the loss of radicals by self-reaction. A radical may transfer an electron or inject a positive hole into a colloidal particle. The advantage of using radicals for the production of excess charge carriers in colloids lies in the fact that carriers of opposite sign are not produced simultaneously. In flash photolysis studies light is absorbed by the colloidal particles. In the case of semiconductor particles, electron-hole pairs are formed and the charge carriers rapidly move to the surface where they may initiate reactions with dissolved compounds, solvent molecules or surface atoms of the colloid. These reactions can again be followed by recording the accompanying changes in optical absorption.

Fluorescence measurements also are quite useful for investigating the behavior of charge carriers in semiconductor colloids. The fluorescence is emitted upon the recombination of charge carriers. Shortly after light absorption, the electron is in the lowest level of the conduction band, and the positive hole in the highest level of the valence band. When they recombine, light with a photon energy equal to the band gap energy is emitted, the lifetime of this fluorescence generally being very short (< 1 ns). Longer decay times and longer wavelengths of the emitted light are observed when trapped charge carriers recombine. The fluorescence is often quenched by solutes at low concentrations. Fluorescence quenching experiments therefore give information about the interaction of the charge carriers with compounds in solution [8].

A precursor of the studies on electron transfer reactions between short-lived radicals and colloidal particles was the development of a fast pulse radiolysis method to measure the polarograms of radicals in the 10^{-5} s range [9]. After considerable information had been acquired about the electron transfer reactions of a few dozen radicals at the mercury electrode, this compact electrode was replaced by metal colloids [10-13], and, somewhat later, by semiconductor colloids [8,14]. These studies led to the detection of the electron-storing properties of certain colloids and of reactions of the stored electrons.

2 Colloidal Metals

2.1 General Remarks

The catalysis of redox processes by colloidal metals was recognized in the early days of colloid chemistry, the decomposition of hydrogen peroxide, the oxidation of carbon monoxide and the hydrogenation of organic compounds on platinum being the most famous examples [15,16]. An important step in the understanding of these catalytic effects were the experiments of Spiro [17,18], in which evidence was put forward that heterogeneous catalysis by metals occurs largely by an electron transfer mechanism. In these experiments, the platinum catalyst was simply used as a foil immersed in the solution. For example, the oxidation of Hg(I) by Ce(IV) ions is strongly catalysed.

The catalysis of reactions of short-lived species, such as free radicals generated by radiation, became of interest in more recent times. When the species whose reactions are to be catalysed is short-lived, the metal has to be present in extremely fine dispersion in order to make the species react with the catalyst before deactivation by another process occurs. Another modern problem is the size dependence of the electrochemical and catalytic properties of the small metal particles.

2.2 Catalysis of Free Radical Reactions by Colloidal Metals

Free radicals generally undergo one-electron transfer processes in homogeneous solution. Two-electron transfer processes, in which two radicals participate, are often highly exoergic. Typical examples are

$$2\,(CH_3)_2\dot{C}OH \rightarrow 2\,(CH_3)_2CO + H_2 \qquad \Delta G = -3.0\ eV \qquad (1)$$

$$2\,\dot{C}O_2^- + 2\,H_2O \rightarrow 2\,CO_2 + H_2 + 2\,OH^- \qquad \Delta G = -3.6\ eV \qquad (2)$$

They do not occur in homogeneous solution, as two radicals encountering each other dimerize or disproportionate. However, the formation of H_2 by short-lived radicals can be catalysed by colloidal metals [10-12]. The most detailed investigation has been carried out with colloidal silver[10,13,19-23]. Figure 1 shows the mechanism in a schematic manner. A solution of colloidal silver (overall concentration: a few 10^{-4} M; particle size: ~7 nm) is exposed to γ-radiation or UV-light. The solution contains an organic solute which decomposes under irradiation to yield free radicals. For example, 1-hydroxy-1-methyl ethyl radicals are produced by both γ-rays and UV-light in solutions containing 0.1 M propanol-2 and 0.05 M acetone[1].

Each radical transfers an electron to the colloidal particles. One colloidal particle can accept and store a large number of electrons until the aqueous solvent is reduced. Under stationary conditions, a certain number x of electrons reside on a particle, thus producing a negative potential sufficiently high for H_2 evolution (n = agglomeration number):

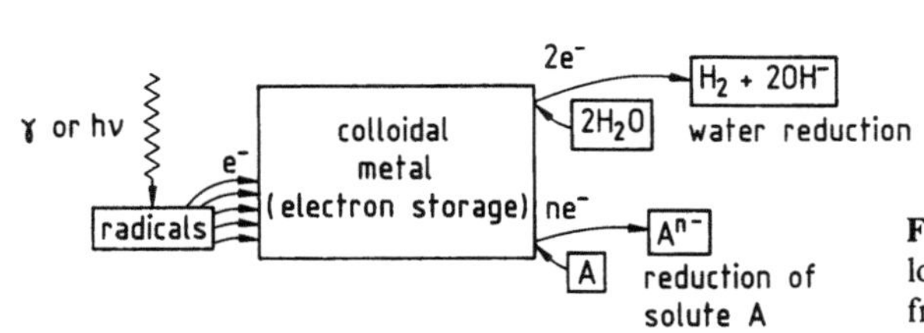

Fig. 1. Schematic description of colloidal metal-catalysed reductions by free radicals

[1] In the case of UV-light, the reactions are: $(CH_3)_2CO + h\nu \rightarrow (CH_3)_2CO^*$; $(CH_3)_2CO^* + (CH_3)_2CHOH \rightarrow 2\,(CH_3)_2\dot{C}OH$. In the case of γ-rays, hydrated electrons, H-atoms, and OH radicals are formed by decomposition of the solvent. The hydrated electrons react with acetone: $e^-_{aq} + (CH_3)_2CO + H^+ \rightarrow (CH_3)_2\dot{C}OH$, and the H and OH radicals with propanol-2: $H(OH) + (CH_3)_2CHOH \rightarrow H_2(H_2O) + (CH_3)_2\dot{C}OH$.

$$2\,(CH_3)_2COH + Ag_n^{x-} \rightarrow 2\,(CH_3)_2CO + 2\,H^+ + Ag_n^{(x+2)-} \tag{3}$$

$$Ag_n^{(x+2)-} + 2\,H^+ \rightarrow Ag_n^{x-} + H_2 \tag{4}$$

$$2\,(CH_3)_2COH \rightarrow 2\,(CH_3)_2CO + H_2 \tag{1}$$

It is recognized, that the overall process consists of the catalysis of reaction 1, an intermediate step being the reduction of water. Reaction 4 can be treated as hydrogen evolution on a compact silver electrode at negative potential. Silver has a small overpotential for H_2 evolution. Of the two steps of which reaction 4 is composed

$$Ag_n^{(x+2)-} + H^+ \rightarrow (Ag_nH)^{(x+1)-} \tag{5}$$

$$(Ag_nH)^{(x+1)-} + H^+ \rightarrow Ag_n^{x-} + H_2 \tag{6}$$

reaction 5 is generally rate determining.

Methods have been worked out for studying the sequence of elementary processes 3 and 4 in detail: The rate of electron transfer from the radicals to the colloidal particles was studied using the method of conductometric pulse radiolysis; the number x of electrons stored in the stationary state (which depends on the rate of radical formation and the pH) was measured, and the lifetime of the stored electrons with respect to reaction 4 was determined. Further, the chemical potential of the stored electrons was ascertained [20]. Figure 2 shows the conductivity changes of a silver sol in which $(CH_3)_2COH$ radicals were produced by continuous γ-irradiation. The mean diameter of the silver particles was 7 nm. The conductivity is expressed as the charge stored per liter of solution, the charge carriers mainly producing the increase in conductivity being the free hydrogen ions which are formed in reaction 3. The conductivity increases until a stationary value is reached. The greater the H^+ concentration of the solution

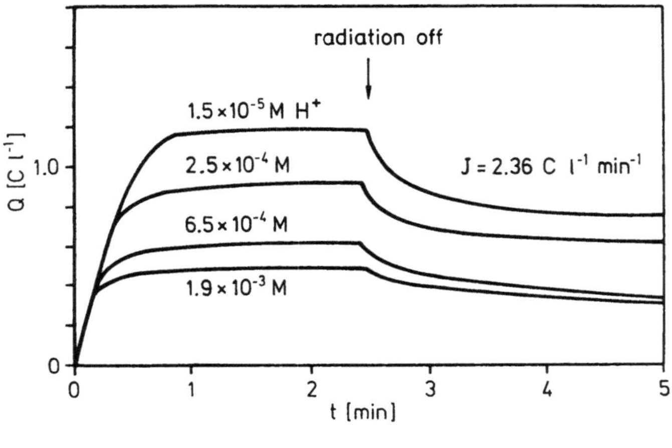

Fig. 2. Stored charge (as calculated from conductivity changes) as a function of irradiation time. Concentration of the silver sol: 2.5×10^{-4} M. J: rate of charging the microelectrode (i.e. rate of radical generation divided by Faraday's equivalent) [20]

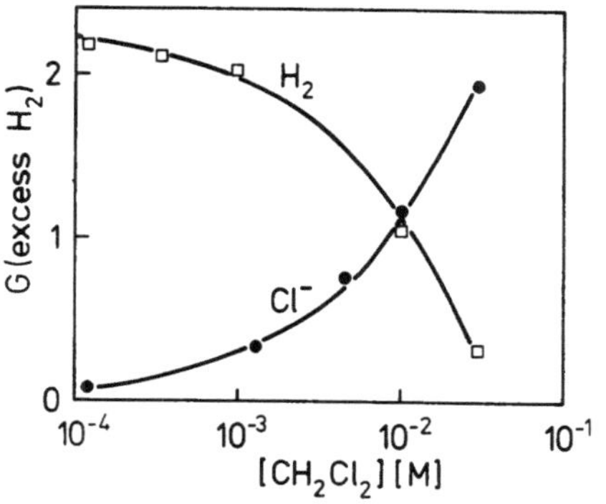

Fig. 3. Irradiation of a silver sol containing propan-ol-2 (0.3 M) and acetone (0.2 M) in the presence of methylene dichloride. H_2 and Cl^- yields as functions of the CH_2Cl_2 concentration [13]

the smaller this stationary value. The more electrons are stored the faster is the discharge reaction 4. In the stationary state, the rates of reactions 3 and 4 are equal. As soon as the irradiation is stopped the conductivity decreases. This decrease is attributed solely to reaction 4. It is seen that 1 coulomb of electrons can readily be stored per liter, each colloidal particle carrying about 500 excess electrons in the stationary state at the lowest H^+ concentration in Fig. 2. The electrons are on a potential of about -1.0 V (the uncharged silver particles having a Fermi potential close to zero on the standard electrochemical scale). The colloidal solution may be characterized by an electrical capacity for the storage of electrons. It amounts to about 1 Farad per liter.

When a reducible compound (that is not directly attacked by the free radicals) is present in the solution, the stored electrons may react with it. Stored electrons may be transferred pairwise to the solute. The reduction of the aqueous solvent and of the solute compete with each other. A typical example is shown in Fig. 3. Methylene dichloride was the solute here, the product of its reduction being the Cl^- ion. It is seen that the H_2-yield decreases as that of Cl^- increases with increasing CH_2Cl_2 concentration. For each H_2 molecule not formed, one Cl^- anion is produced. As two electrons are necessary to produce one H_2 molecule, one has to conclude that methylene dichloride reacts with stored electrons in a two-electron transfer process:

$$2\,e^- + CH_2Cl_2 \xrightarrow{H_2O} Cl^- + CH_3Cl + OH^- \qquad (7)$$

The methyl chloride formed could also be detected [13]. Many other examples for multielectron transfer from the charged silver microelectrode have been found. NO_3^- is reduced to NH_3 in an 8-electron transfer, Cd^{2+} to Cd^0 in a two-electron transfer [22]. One-electron transfer reactions, which do not take place in homogeneous solution, can also be catalysed. For example, many organic radicals are not able to reduce Ag^+ or Tl^+ in solution, the reason being that the potentials of the redox systems involved are very negative ($E^0(Ag^+/Ag^0) = -1.8$ V; $E^0(Tl^+/Tl^0) = -1.9$ V; Ag^0 and Tl^0 being free atoms in solution!). In the presence of colloidal silver, the products are Ag or Tl atoms strongly bound to the surface of the microelectrode, the binding energy facilitating the reduction [19]. Not all organic radicals are able to charge a microelectrode. The more electropositive radicals may even discharge a microelectrode.

Colloids of more electronegative metals such as cadmium [24] and thallium [25] also act as catalysts for the reduction of water. In the colloidal solution of such a metal, an appreciable concentration of metal ions is present. The transferred electrons are first used to reduce the metal ions, thus bringing the Fermi potential of the colloidal particles to more negative values. After all the metal ions have been reduced, excess electrons are stored as in the case of silver.

Colloids of metals with practically zero overpotential for H_2 evolution such as platinum [21] and iridium [26], are very efficient at catalysing H_2 production. The transferred electrons are immediately converted into adsorbed H atoms which then undergo the Heyrovsky reaction to form H_2: $e^- + H + H^+ \rightarrow H_2$. The gold microelectrode takes an intermediate position between those of Ag and Pt, the stationary electron concentration being smaller than in the case of the silver microelectrode [27,28]. H_2 evolution occurs less efficiently in alkaline solutions (pH > 10), since the rate of the Heyrovsky reaction becomes too slow. Unter these conditions, a colloid catalysed disproportionation of the radicals occurs.

The radicals in Eqs. (1) and (2) have strong reducing power, and the reverse reactions (i.e. the one-electron reduction of acetone or CO_2) hardly occur. In the case of radicals of less negative potential, the back reaction has to be taken into consideration. The most frequently studied radical of low reduction potential is the half-reduced form of methyl viologen (1,1'-dimethyl-4,4'-bipyridylium dichloride, MV^{2+}). This compound has been proposed as an electron transfer relay in systems for catalysing the reduction of water by visible light [29]. In these systems, a light absorbing sensitizer such as $Ru(bpy)_3^{2+}$ is excited and transfers an electron to MV^{2+} to produce the radical cation MV^+. This radical is long-lived as its disproportionation is an endoergic reaction. In the presence of platinum it reduces water:

$$2\,MV^+ + 2\,H_2O \xrightarrow{Pt} 2\,MV^{2+} + H_2 + 2\,OH^- \tag{8}$$

In an irradiated Pt-sol containing the MV^{2+} relay, a microelectrode $Pt(H^+/H_2)$ is in contact with the redox system MV^{2+}/MV^+ in solution. In principle, the reactions occurring in such a solution can be dealt with on the basis of corrosion theory [30] as recently modified by Spiro [18]. The colloidal particles attain a mixed potential in the stationary state. The anodic and cathodic currents at this potential can be calculated, and these currents lead directly to the catalytic rate [31,32]. Iridium also is an efficient catalyst for the reduction of water by MV^+ [33].

A detailed treatment of the transport and kinetics at microheterogeneous electrodes was given by Albery and Bartlett [34]. The electron transfer of MV^+ to Pt-particles can be described by a bimolecular rate constant k_2:

$$\frac{1}{k_2} = \frac{1}{4\pi R^2 L}\left[\frac{1}{k'} + \frac{1}{(Dk_1)^{1/2} + D/R}\right] \tag{9}$$

where $L = 6.02 \times 10^{23}$ molecules/mol, R is the radius of the colloidal particles, D the diffusion coefficient of MV^+, and k_1 a pseudo-first order rate constant for possible competing reactions of MV^+ in homogeneous solution. k' is the electrochemical rate constant in $cm \cdot s^{-1}$. Note that if k' is large and k_1 small, the usual expression

$$k_2 = 4\pi RDL \tag{10}$$

for a diffusion-controlled bimolecular reaction is obtained. Reaction 8 was studied by stopped flow spectrophotometry. As could be expected, the rate of reaction depended on the gaseous pretreatment of the Pt-particles [35,36].

The MV^+/Pt system was also studied in a number of radiation chemical experiments, in which a wide range of concentrations of all the participating species was covered [37–41]. These investigations also revealed that MV^{2+} cannot be used as a relay system for practical purposes in which a great turn-over number for the conversion $MV^{2+} \rightarrow MV^+ \rightarrow MV^{2+}$ is required. The reason lies in the fact that MV^+ is catalytically reduced in a side reaction to yield various products.

Adsorbed hydrogen atoms which are the intermediates of reaction 8 can be used for the hydrogenation of ethylene and acetylene. The light driven reaction occurs according to the following scheme:

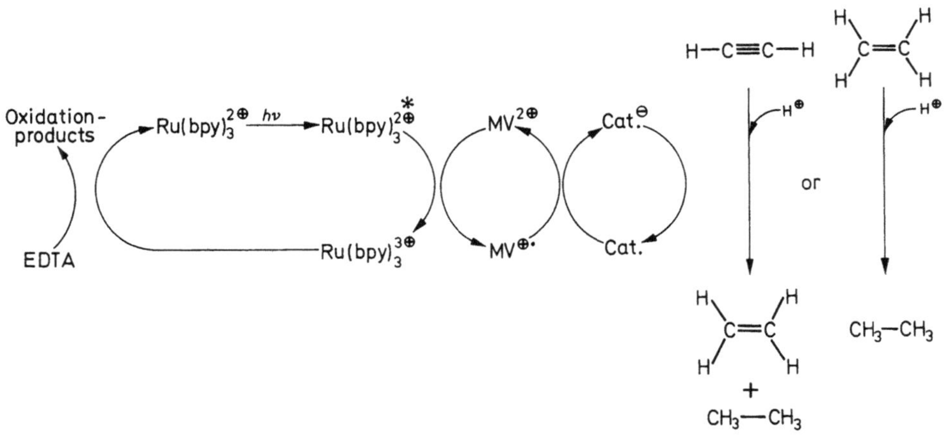

$Ru(bipy)_3^{2+}$ is the sensitizer which transfers an electron to the MV^{2+} relais. The $Ru(bipy)_3^{3+}$ formed in this reaction is reduced by the sacrificial electron donor sodium ethylenediaminetetra-acetic acid, EDTA. Cat is the colloidal catalyst. With platinum, the quantum yield of C_2H_4 hydrogenation was 9.9×10^{-2}. The yield for C_2H_2 hydrogenation was much lower. However, it could substantially be improved by using a Pt colloid which was covered by palladium [42]. This example demonstrates that complex colloidal metal catalysts may have specific actions. Bimetalic alloys of high specific area often can be prepared by radiolytic reduction of metal ions [43,44].

Reactions of oxidizing radicals with colloidal metals have been investigated less thoroughly. OH radicals react with colloidal platinum to form a thin oxide layer which increases the optical absorbance in the UV and protects the colloid from further radical attack [45]. Complexed halide atoms[2], such as Cl_2^-, Br_2^-, and I_2^-, also react

[2] The complexed halide atoms are produced by high energy radiation in solutions of colloids that contain halide anions X^- and are saturated with nitrous oxide. Hydrated electrons formed in the radiolysis of the aqueous solvent react with N_2O according to $N_2O + e_{aq}^- + H_2O \rightarrow N_2 + OH^- + OH$ to form additional OH radicals. Ions X^- are oxidized by OH, the atoms X thus formed react rapidly with X^- to yield X_2^- radicals.

with colloidal platinum. The first step in this reaction consists of the transfer of a halogen atom to platinum. The chemisorbed atom then may undergo hydrolysis. In the case of Br_2^-, the equilibrium

$$Pt_n{-}Br \underset{-H_2O}{\overset{H_2O}{\rightleftarrows}} Pt_nOH + H^+ + Br^- \tag{11}$$

could be observed, the equilibrium constant being 1×10^{-6} M^2.

2.3 Size Dependence of the Potential of a Microelectrode

The standard equilibrium potential of a metal/metal ion electrode is shifted to cathodic potentials with decreasing particle size. Taking a compact metal electrode as reference, the change in potential for a microelectrode of radius r is

$$\Delta\epsilon = -\frac{2M}{zF\varrho}\frac{\gamma}{r} \tag{12}$$

where γ is the surface free energy of the metal, M the molar mass, ϱ the specific mass, z the number of charges on a metal ion, and F Faraday's constant. This formula may be used to calculate the potential of microelectrodes down to 2 nm [46]. For smaller particles new effects determine $\Delta\epsilon$ as a transition from metal to molecules takes place. Very little is known about the dependence of the equilibrium potential of systems

$$M_n \rightleftarrows M_{n-1} + M^{z+} + ze^- \tag{13}$$

on the agglomeration number n of the microelectrode (M: metal).

Figure 4 shows what is known to date about the standard potential of the silver microelectrode as a function of the agglomeration number n [4]. For n = 1, i.e. the redox system

$$Ag^0 \rightleftarrows Ag^+ + e^- \tag{14}$$

where Ag^0 is an isolated free silver atom in aqueous solution, one calculates a potential of —1.8 V by subtracting $\Delta G_{sub}/e$ from the standard potential of 0.79 V of the compact

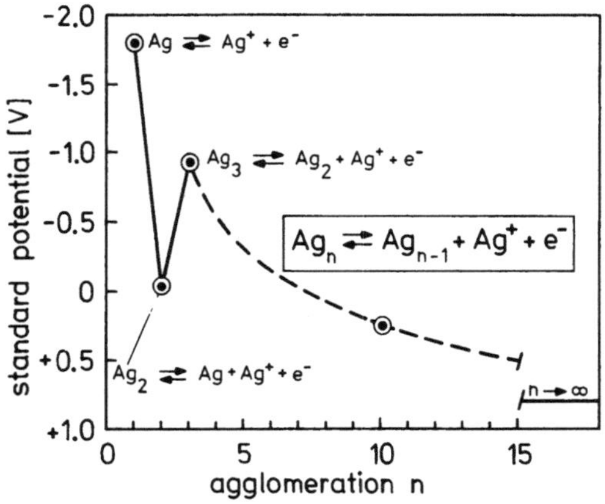

Fig. 4. Standard potential of the silver microelectrode as a function of the agglomeration number n [4]

silver electrode [47]. ΔG_{sub} is the free energy of sublimation of silver at room temperature, which amounts to 2.60 eV. The hydration energy of the silver atom was neglected in this calculation. Since the hydration free energies of atoms (for example of the heavier noble gas atoms) are less than 0.1 eV, one may regard the above potential of -1.8 V for the Ag^+/Ag^0 system as a reasonable approximation. Silver atoms can be produced in pulse radiolysis experiments in which a silver salt solution is irradiated. The hydrated electrons react with silver ions according to:

$$Ag^+ + e^-_{aq} \rightarrow Ag^0 \tag{15}$$

The absorption spectrum of Ag^0 in aqueous solution has a broad band with a maximum at 340 nm. The Ag^0 atoms rapidly disappear as they react with Ag^+ ions to form Ag_2^+ which absorbs at 310 nm

$$Ag^0 + Ag^+ \rightarrow Ag_2^+ \tag{16}$$

In the presence of substances that react with Ag^0, the 340 nm absorption decays more rapidly and the rate constant of reaction can be calculated from this decay. It was found in this way that free silver atoms are indeed a strongly reducing species. They reduce Fe^{3+} to Fe^{2+} and Cu^{2+} to Cu^+ (note that these reactions would not occur with the silver atoms at the surface of a compact electrode!); organic compounds containing electrophilic groups such as $ClCH_2COOH$ or CH_3NO_2 are reduced by Ag^0 via electron transfer [48].

Ag_2 and Ag_3 are present in small concentrations in silver vapor [49]: Mass spectrometric determinations of their equilibrium concentrations made it possible to calculate their thermodynamic properties in the gas phase. Using these properties and the potential of -1.8 V for the Ag^+/Ag^0 system, the standard potentials of the silver redox systems in Fig. 4 with $n = 2$ and $n = 3$ were calculated. Again, the hydration energies of the species involved were neglected, the corresponding error in the calculated potentials probably not exceeding 0.1 V. It is seen that the standard redox potential shows some oscillatory behavior. Unfortunately, no experimental data on the thermodynamic properties of particles with $n > 3$ are available to enable one to draw the curve of potential vs. n in a reasonable manner for $n \gg 3$. Therefore the curve in Fig. 4 was dashed for $n > 3$, assuming that the potential will somehow tend towards 0.79 V for $n \rightarrow \infty$. The point for $n = 10$ was obtained from the experiments with the growing silver microelectrode described below. Further oscillations may still exist in the dashed part of the curve.

There are no known methods of preparing Ag_n clusters with small values of n as reasonably stable species in aqueous solution and studying their electrochemical behavior. However, the method of the growing silver microelectrode [50] allows one to obtain some information about the electrochemical properties of small silver particles. As mentioned above, silver atoms Ag^0 can be produced at a known concentration in a solution of a silver salt by a short pulse of high energy radiation. The silver atoms form Ag_2^+, and the Ag_2^+ particles form larger and larger products. The mechanism of growth of silver clusters, i.e. of the process $Ag_n + Ag_m \rightarrow Ag_{n+m}$, is not known in detail. It may be an electrochemical mechanism: $Ag_n + Ag_m \rightarrow nAg^+ + Ag_m{}^{n-}$, followed by $Ag_m{}^{n-} + nAg^+ \rightarrow Ag_{m+n}$. If $n < m$, the potential in Ag_n is more negative than in Ag_m (see eq. 12), i.e. there exists a driving force for the transfer of

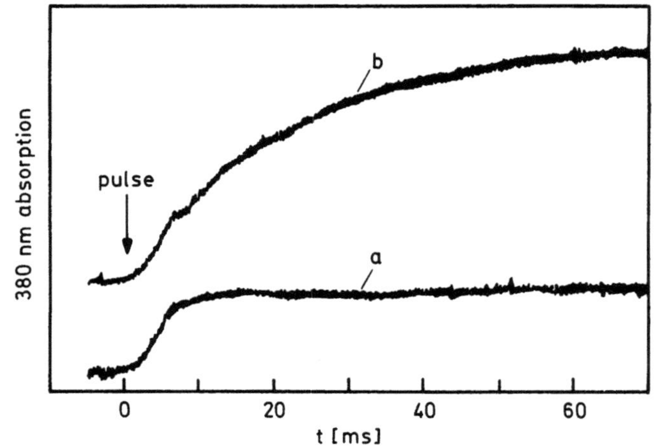

Fig. 5. Oscillographic traces of the 380 nm absorption of growing silver particles [50]: (a) only Ag atoms were produced (5×10^{-7} M); (b) 5×10^{-7} M Ag atoms and 1.5×10^{-6} M Cu^+ ions were produced initially

electrons from Ag_n to Ag_m. These processes can be followed by recording the accompanying changes in optical absorption. The final product is a silver colloid with a particle agglomeration number of some thousands. Colloidal silver has a narrow absorption band at 380 nm which is caused by plasma excitation of the electrons in the metal. Oligomeric silver molecules, which do not possess metal character, do not show this absorption band. The question may first be asked how many silver atoms have to be in a particle to bring about this metal property? With respect to the pulse experiment, one may ask at which size of the growing silver particles is the 380 nm absorption developed?

Curve (*a*) in Fig. 5 shows that the 380 nm absorption reaches its full intensity about 10 ms after the pulse of radiation. The kinetic analysis showed that the main particle size was 12 Ag atoms/particle at this time. However, the particles probably contained some additional Ag^+ ions, the true particle size therefore being slightly greater. The only difference with respect to the absorption of larger colloidal particles consisted of a larger width of the 380 nm absorption band. This width actually became narrower at longer times, which shows that the particles continued to grow at times in excess of 10 ms after the pulse. During the agglomerations, the redox potential of the growing microelectrode will move from an initial —1.8 V to a final value close to 0.79 V. A redox reaction in which the growing microelectrode, i.e. the system $Ag_n \rightleftarrows Ag_{n-1} + Ag^+ + e^-$, is involved can occur only after the appropriate particle size has been reached. This is demonstrated in the experiment of curve b in Fig. 5. A solution containing silver and copper ions was pulsed. It is known from radiation chemistry that irradiation of such a solution produces both Ag^0 and Cu^+, the relative concentration ratio between these species being dependent on the Ag^+/Cu^{2+} ratio. The concentrations and the dose in the pulse were chosen such as to produce as many silver atoms as in the experiment of curve a, and three times the amount of Cu^+ ions. The first step in the curve is identical to the build-up of the 380 nm absorption in curve (*a*). The second step starts shortly before the first step has reached its plateau at $n = 12$. The additional formation of 380 nm absorption, i.e. of formation of additional silver, is due to the catalysed reaction

$$Cu^{+} + Ag^{+} + Ag_{n} \rightarrow Cu^{2+} + Ag_{n+1} \quad (17)$$

where Ag_n represents the growing silver particles. Cu^+ cannot produce free silver atoms in solution, the potential of the Cu^{2+}/Cu^+ system (0.2 V) being much more positive than that of the Ag^+/Ag^0 system (—1.8 V). At first the Cu^+ ions are idle spectators of the agglomeration of the initially generated Ag^0 atoms, but as soon as the potential of the growing particles attains the potential of the Cu^{2+}/Cu^+ system the additional reduction of Ag^+ according to Eq. (17) can start. Detailed analysis showed that this is reached at $n = 10$.

Figure 4 illustrates that our present knowledge about the dependence of the standard potential of very small silver microelectrodes on the agglomeration number is rather fragmentary. Even less is known about this dependence for other metals. The experiments of Fig. 5 prove that the rate of an electrochemical reaction in which a small microelectrode is involved, may strongly depend on the size of the microelectrode.

3 Colloidal Semiconductors: Metal Sulfides

3.1 General Remarks

The initiation of photochemical reactions on semiconductor particles has, in priniciple, been familiar for a long time, the photographic process being the most famous example. Reactions on inorganic sulfides were also recognized several decades ago. For example, ZnS, a component of the white pigment lithopone, was shown not only to be photolysed by UV light but also to catalyse the photodecomposition of water [51,52].

The chemical reactions that are catalysed by the smaller colloidal particles are generally the same ones that occur on the larger suspended particles. In work on colloidal systems, however, kinetic studies can often be carried out in a more detailed manner. As the fundamental aspects of reactions on colloidal particles are the main object of the present article, the reactions on larger particles have only been referred to occasionally. The fact that the reactions on colloidal particles and larger suspended particles are quite similar is somewhat surprising at first sight, since electrons and positive holes should move differently in both cases. In a compact semiconductor electrode in contact with a redox system in solution, the electronic energy levels are bent upwards close to the surface in the case of a n-semiconductor, and downwards in the case of a p-semiconductor. This band bending is of great importance for the charge separation in the photoelectrochemistry of semiconductors [53–55]. The thickness of the depletion layer in which the energy levels are bent is of the order of 100 nm. An electron produced by light absorption in the depletion layer of an n-semiconductor is driven into the interior, while a positive hole migrates to the surface. In the larger particles, the dimensions of which substantially exceed 100 nm, similar charge separation effects should occur. However, during the first stages of illumination band flattening may take place. In the case of an n-semiconductor, the electrons generated by the first absorbed photons move into the interior while the holes move to the surface or are even transferred into the solution. This would lead to a photostationary state in which the original band bending is offset by the excess electrons on the particles. In colloidal particles of diameters much smaller than 100 nm, no band bending exists.

At first sight, one would predict that a great percentage of the charge carriers generated recombine and that the photochemical yields of the products would be very low. However, cases are known in which the yields are close to 100%. Moreover, reactions are sometimes catalysed by colloidal particles in which two electrons or two positive holes are delivered to a substrate. The time between two successive photon absorptions being of the order of milliseconds under the usual illumination conditions, one recognizes that the two electrons or holes are not produced simultaneously. In the chapter on colloidal metals, a mechanism for the storage of electrons was described, i.e. for the reduction of substrates via multielectron transfer, and similarly, one has to keep in mind the question how electrons or holes (or, more generally, how reducing and oxidizing equivalents) may be stored in semiconductor particles.

3.2 Photoanodic Dissolution of Colloidal Semiconductors

When colloids of metal sulfides, selenides, phosphides and arsenides are illuminated in the presence of air, decomposition takes place. Metal ions migrate into solution and sulfate, selenite, phosphate or arsenate are formed. The process was first investigated in the case of colloidal CdS [8]:

$$CdS + 2\,O_2 \xrightarrow{\text{light}} Cd^{2+} + SO_4^{2-} \quad (18)$$

When the analysis is carried out shortly after the illumination, small amounts of sulfite can also be detected.

The photoanodic dissolution also occurs in the electrochemistry and photoelectrochemistry of compact electrodes of these materials. In fact, it is the most serious obstacle to the practical use of semiconductors such as CdS in photoelectrochemical cells [55]. The product of corrosion in the absence of oxygen is sulfur. In the presence of oxygen, sulfate ions are formed as in the case of the colloidal particles [56].

Although it presents an obstacle in practical applications, the photoanodic corrosion of colloids has often been used to obtain information about the interaction of dissolved compounds with the photo-produced charge carriers, as it was found that solutes can influence the rate of the dissolution. Both promoting and retarding effects were observed [8]. The rate of dissolution is readily followed by recording the decrease in the intensity of the absorption spectrum of the colloid upon illumination, or more precisely, by determining the yields of metal and sulfate ions in solution.

Figures 6 and 7 show the absorption spectra of colloidal CdS and ZnS at various times of illumination. The two colloids were prepared by adding an NaSH solution to solutions of $Cd(ClO_4)_2$ or $Zn(ClO_4)_2$, respectively, colloidal silicon dioxide (commercially available from Dupont: Ludox HS30) being present at 6×10^{-3} M as stabilizer in both cases. The absorption starts in both cases close to the wavelengths that correspond to the photon energies (515 nm or 2.4 eV for CdS; 340 nm or 3.7 eV for ZnS) at which the absorptions begin in the macrocrystalline materials. It is seen that illumination causes not only a decrease in the intensity of the absorption spectrum but also a change in the shape of the spectrum. The onset of light absorption is shifted towards

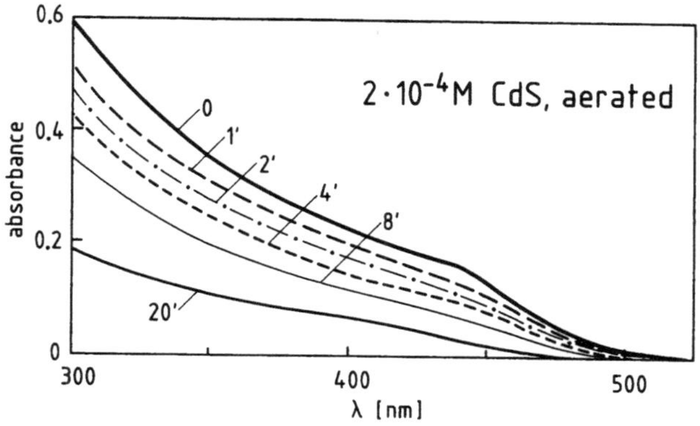

Fig. 6. Absorption spectrum of $2 \cdot 10^{-4}$ M CdS at various times of illumination ($\lambda > 490$ nm) in the presence of air [8]) Stabilizer: $6 \cdot 10^{-3}$ M colloidal SiO_2 (Ludox HS30). Initial mean diameter of particles: 3 nm

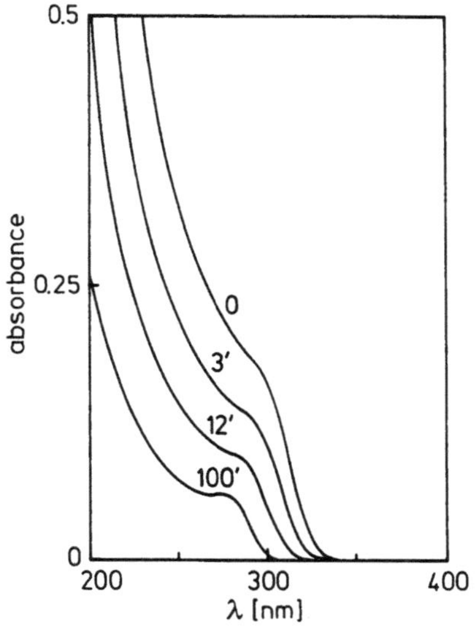

Fig. 7. Absorption spectrum of $2 \cdot 10^{-4}$ M ZnS at various times of illumination ($\lambda > 280$ nm) in the presence of air [70]. Initial mean diameter of the particles: 3 nm

shorter wavelengths. The colloidal particles become smaller during the photoanodic dissolution. The observation of changes in the shape of the absorption spectrum was one of the first indications of the occurrence of size quantization effects in extremely small particles. These effects are described further in Sect. 5.1.

The photoanodic dissolution of sulfides is described for CdS by the following equations:

$$CdS \xrightarrow{h\nu} CdS(e^- + h^+) \tag{19}$$

$$e^- + O_2 \longrightarrow O_2^- \tag{20}$$

$$h^+ + CdS \longrightarrow Cd^{2+}S^- \tag{21}$$

$$Cd^{2+}S^- + O_2 + O_2^- \longrightarrow Cd^{2+} + SO_4^{2-} \tag{22}$$

Absorption of a light quantum leads to an electron-hole pair Eq. (19). The electron reacts with an adsorbed oxygen molecule Eq. (20), and the hole semi-oxidizes a sulfide anion at the surface Eq. (21). Further oxidation of the sulfide anion occurs by O_2^- and O_2 Eq. (22). The number of Cd^{2+} ions formed equals that of the sulfate anions [57]. The oxidation of illuminated CdS powders was investigated by measuring the O_2 consumption and by detecting the superoxide radical, O_2^-, by an ESR spin trapping method [58, 59].

The photochemical yield in the experiments of Fig. 6 is 0.04 CdS molecules dissolved per photon absorbed. The yield itself is not constant during the illumination but decreases as the degree of dissolution increases, i.e. with decreasing particle size.

In the absence of oxygen, CdS is not photo-dissolved. The electrons in the conduction band of CdS are on a potential of —0.6 to —0.9 V (depending on the pH) and should be able to reduce water:

$$2\,H_2O + 2\,e^- \rightarrow H_2 + 2\,OH^- \tag{23}$$

However, this process hardly occurs, recombination being too fast in the absence of O_2.

Methyl viologen (1,1′-dimethyl-4,4′-bipyridylium dichloride, MV^{2+}) promotes photoanodic dissolution in aerated CdS solution [60]. Figure 8 shows how the rate of dissolution depends on the MV^{2+} concentration. The colloid has a weak fluorescence at 620 nm which is quenched by MV^{2+}. The curves for fluorescence and dissolution in Fig. 8 are symmetric, which indicates that the two processes have a common intermediate that reacts with MV^{2+}. These effects are explained by the following mechanism:

$$e^- + MV^{2+} \rightarrow MV^+ \tag{24}$$

$$MV^+ + O_2 \rightarrow MV^{2+} + O_2^- \tag{25}$$

$$h^+ + CdS \rightarrow Cd^{2+}S^- \xrightarrow{O_2,\,O_2^-} Cd^{2+} + SO_4^{2-} \tag{26}$$

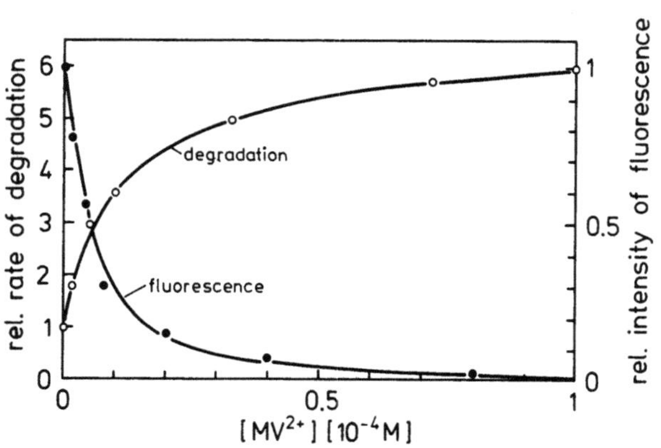

Fig. 8. Intensity of fluorescence and rate of photoanodic dissolution of colloidal CdS as function of the concentration of added methyl viologen [60]

As MV^{2+} is strongly adsorbed at colloidal CdS, it efficiently scavenges the electrons formed by light absorption. The semi-reduced methyl viologen is reoxidized by O_2, and the positive holes have a greater chance of oxidizing sulfide anions. A strong promotion of the photoanodic dissolution was also found for Tl^+ ions. Adsorbed Tl^+ is reduced by e^- to an adsorbed Tl atom which then is reoxidized by O_2. A weaker promoting effect was observed for nitrobenzene. Intermediate formation of $C_6H_5NO_2^-$ and the reoxidation of this anion by O_2 explains the effect. All these promoters are practically ineffective when oxygen is not present in the solution [8]. The photoanodic dissolution is retarded when small amounts of excess NaSH are present. Illumination leads to sulfur and sulfate. After all the excess SH^- ions have been oxidized, the dissolution of the colloid takes place [8].

ZnS in aerated solution is photo-dissolved much more rapidly than CdS, the quantum yield being 0.32 ZnS molecules consumed per photon absorbed [61]. This is possibly due to a stronger adsorption of O_2 on the colloidal ZnS particles. In the absence of oxygen, a very slow photoanodic dissolution was observed, the main products being H_2 and sulfur (quantum yield of H_2: 2×10^{-3} molecules/photon absorbed). The formation of H_2 is explained by the mechanism of Equs. 19 and 23 formulated for CdS. That this reaction can occur in ZnS is understood in terms of the more negative potential of the electrons in the conduction band of ZnS (~ -1.4 V vs. standard hydrogen electrode) which provides more driving force for the H_2 evolution than in the case of CdS.

Photoanodic dissolution in the presence of air and its promotion by methyl viologen was also observed for alkaline solutions of colloidal cadmium phosphide, Cd_3P_2 [62], and cadmium arsenide, Cd_3As_2 [63]. Bismuth sulfide, Sb_2S_3, photo-dissolves in the presence of air mainly according to

$$3/2\ O_2 + Sb_2S_3 \xrightarrow{\text{light}} Sb_2O_3 + 3/8\ S_8 \qquad (27)$$

Small amounts of SO_3^{2-} and SO_4^{2-} also are formed. The acceleration of reaction 27 by a factor of 24 in the presence of 2×10^{-5} M MV^{2+} is remarkably strong [64]. The photoanodic dissolution of ZnSe and CdSe in the presence of air leads to selenium and small amounts of selenite. Colloidal selenium also dissolves upon illumination in the presence of air the quantum yield being 0.007 Se atoms per photon absorbed. The only product of corrosion is selenite. In the presence of methyl viologen, the reaction is accelerated.

3.3 Fluorescence of CdS and ZnS

The luminescence of macrocrystalline cadmium and zinc sulfides has been studied very thoroughly [65]. The colloidal solutions of these compounds also fluoresce, the intensity and wavelengths of emission depending on how the colloids were prepared. We will divide the description of the fluorescence phenomena into two parts. In this section we will discuss the fluorescence of larger colloidal particles, i.e. of CdS particles which are yellow as the macrocrystalline material, and of ZnS particles whose absorption spectrum also resembles that of the macrocrystals. These colloids are obtained by precipitating CdS or ZnS in the presence of the silicon dioxide stabilizer mentioned in Sect. 3.2, or in the presence of 10^{-2} M sodium polyphosphate [66,67], or surfactants such as sodium dodecyl sulfate and cetyldimethylbenzyl-ammonium

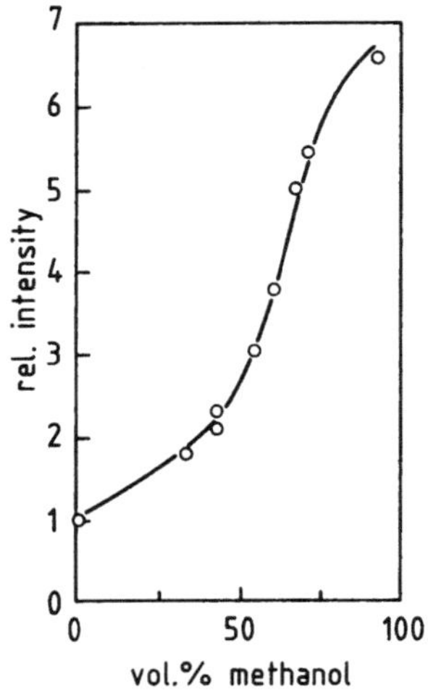

Fig. 9. Fluorescence intensity of a CdS sol in methanol-water mixtures of various compositions [73)]

chloride [68,69)]. The fluorescence of extremely small CdS and ZnS particles, which show size quantization effects, is described in Sect. 5.3.

The first observations on the fluorescence of colloidal CdS were made using a colloid stabilized by colloidal silicon dioxide [8)]. The fluorescence spectrum consisted of a broad band with the maximum between 580 nm and 650 nm. The reproducibility of this red fluorescence was very poor. In the presence of excess Cd^{2+} ions the intensity of the fluorescence was increased, which indicates that anion vacancies were centers of luminescence. Aging of the sol for a few weeks in the dark and in the absence of air was accompanied by an increase in fluorescence intensity by a factor of ten and a gradual red shift of the fluorescence band. However, even after this increase, the fluorescence quantum yield was still below 10^{-3}.

Most striking is the increase in the fluorescence intensity of a CdS colloid as it undergoes photoanodic dissolution. As the colloidal particles become smaller they fluoresce with a greater quantum yield [70)]. Very small CdS particles prepared by the methods described in section 5.1, fluoresce with a quantum yield of 3% [71)].

CdS sols with fluorescence quantum yields close to 100% are obtained by certain surface modification procedures [72–73a)]. For example, dilution of an aqueous sol with alcohol enhances the fluorescence yield as it is shown by Fig. 9. Similarly, CdS fluoresces stronger when it is prepared in acetonitrile-water mixtures [73b)]. Further, the luminescence of colloidal CdS is activated by the addition of either $Ag(ClO_4)$ or colloidal Ag_2S. The degree of activation depends on how strongly CdS and Ag_2S particles are attached to each other. When polyphosphate is used as stabilizer this attachment can be controlled by divalent metal ions such as Ba^{2+} or Mg^{2+} which bind to the phosphate carrier. Addition of NaOH or triethylamine to a CdS sol containing excess Cd^{2+} ions also leads to a substantial increase in fluorescence intensity. In the usual colloids where the quantum yields of fluorescence are only a few percent, the radiationless recombination of the charge carriers is the prevailing process that competes with the fluorescence. Whether the recombination of the charge carriers occurs with or without radiation depends on the nature of the defects states on the surface of the particles where the charge carriers are trapped. The action of methanol could consist of blocking the defect states at which the radiationless recombination occurs. Similarly, the promoting effect of a large excess Cd^{2+} concentration and higher pH could be explained as a protecting cadmium hydroxide layer being formed around the CdS particles [73, 73a].

ZnS and CdS form mixed crystals in the macrocrystalline state over the whole range of composition. Simultaneous precipitation of ZnS and CdS leads to a colloid whose absorption and fluorescence properties indicate that it is not a mixture of the two homo-colloids but a true co-colloid. The onset of absorption and the maximum of the fluorescence band shift gradually as the Cd^{2+} mole fraction varies. When the solutions of the two colloids are prepared separately and then mixed, the absorption and fluorescence spectra represent superpositions of the spectra of the single components. When Cd^{2+} ions are added to the solution of colloidal ZnS, the 428 nm fluorescence of ZnS is strongly quenched. One recognizes from Fig. 10 that the addition of 2×10^{-5} M Cd^{2+} ions to 2×10^{-4} M colloidal ZnS is sufficient for practically 100% quenching. The figure also shows the absorption and fluorescence spectra of solutions that contain larger amounts of added Cd^{2+}. With increasing Cd^{2+} concentration a new fluorescence band appears with the maximum at 580 nm. The explanation can be given on the basis of co-colloid formation. Cd^{2+} ions substitute Zn^{2+} ions at the surface. Although the absorption of light occurs mainly in the unchanged ZnS part of the colloid, the fluorescence is emitted from that part at the surface where Cd^{2+} ions have penetrated and formed the co-colloid [70].

CdS sols of 5 nm particle size stabilized by 10^{-2} M sodium hexametaphosphate show a weak green fluorescence ($\lambda_{max} \sim 515$ nm) besides the red fluorescence at longer wavelengths. The red fluorescence was ascribed to the recombination of charge carriers at anion vacancies [67]. The green fluorescence arises from the recombination of free charge carriers. The red fluorescence is quenched by very small amounts of

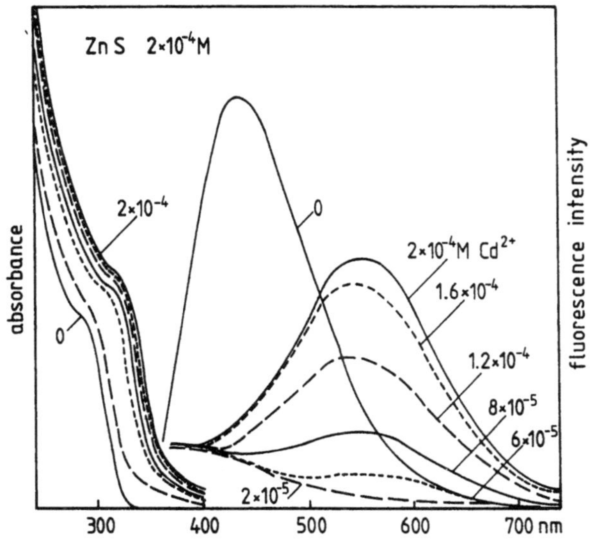

Fig. 10. Spectrum of absorption and fluorescence of ZnS upon the addition of various amounts of Cd^{2+} [70]

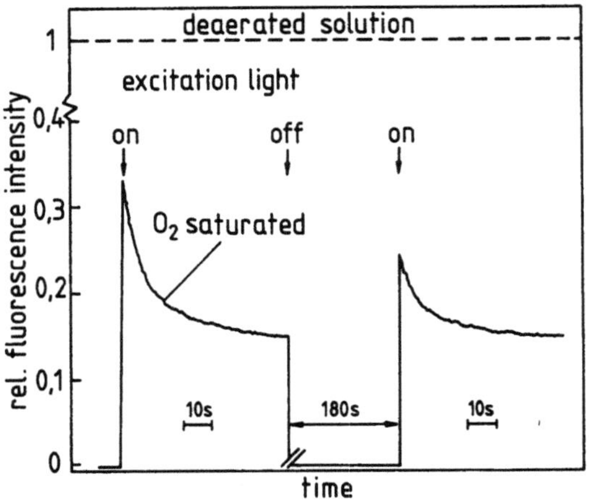

Fig. 11. Time dependence of the fluorescence of ZnS in a solution saturated with oxygen [70]

methyl viologen, one MV^{2+} ion adsorbed on a colloidal particle being sufficient for quenching. MV^{2+} also exerts a striking effect on the green luminescence. Within about 10 minutes of the addition of MV^{2+} a green fluorescence at 530 nm developed. It has been suggested that Cd^{2+} vacancies are responsible for this green emission assuming that adsorbed MV^{2+} is able to create such vacancies. Doping the colloid with Cu^{2+} produced the characteristic emission of this activator at 820 nm[67].

The fluorescences of CdS and ZnS at longer wavelengths, which are ascribed to the recombination of trapped charge carriers, decay in a multiexponential manner in the 100 ns range, the rate becoming slower with decreasing temperature. The longer the wavelength at which the decay is observed, the slower is the decay. Two different explanations have been given for this effect. The first explanation invokes different Franck-Condon energies in the tunneiling of electrons which are trapped in surface states of different energies [70]. In the second explanation there is a correlation between the distance at which the electron and the hole are trapped and the wavelength of light emission because of the Coulomb interaction of the two trapping sites [74].

Oxygen influences the fluorescence of colloidal ZnS in a complex manner. Fig. 11 shows the fluorescence intensity of a solution saturated with oxygen plotted as a function of the time of residence in the fluorimeter. Immediately on switching on the exciting light beam, the fluorescence intensity is about one third of that in the absence of O_2. Within a few seconds the intensity decreases by another factor of 2 or 3. This decrease becomes larger and faster as the intensity of the exciting light beam of the fluorimeter increases. When the sample was left in the dark for 180 s the fluorescence intensity recovered by a factor of about 1.5 but decreased again upon further illumination. However, at much longer illumination times, as in Fig. 11, the fluorescence intensity increased again and finally became even greater than that at the start of the measurements. The fluorimeter not only served as an analytical instrument but also as a source of light for the photochemical modification of the colloid. The following reactions explain these effects:

$$e^- + h^+ \rightarrow h\nu_{fl} \qquad (28)$$

$$e^- + O_{2_{ads}} \rightarrow O^-_{2_{ads}} \qquad (29)$$

$$O^-_{2ads} + e^- + 2\,H^+ \rightarrow H_2O_2 \tag{30}$$

$$h^+ + ZnS \rightarrow Zn^{2+}S^- \xrightarrow{O_2, O_2^-} Zn^{2+} + SO_4^{2-} \tag{31}$$

The recombination of trapped electrons and holes produces the fluorescence. Adsorbed oxygen scavenges electrons producing O_2^- which also is adsorbed. O_2^- is a much better quencher than O_2. Its accumulation under illumination therefore leads to the decrease in fluorescence intensity. During the dark period O_2^- disappears. During the illumination in the presence of oxygen, the colloid undergoes photoanodic dissolution (see Sect. 3.2). The ZnS particles become smaller in this way, and this finally leads to an increase in fluorescence yield as already described for CdS.

3.4 Photocathodic Dissolution and Hydrogen Evolution

In the presence of sulfite anions and absence of oxygen, CdS sols undergo photocathodic dissolution [75]:

$$CdS + SO_3^{2-} + H_2O \rightarrow Cd^0 + SO_4^{2-} \text{ (or } 1/2\ S_2O_6^{2-}) + SH^- + H^+ \tag{32}$$

The metal formed is deposited on the colloidal particles and causes the absorption of the solution to increase at all wavelengths. The deposited cadmium is very reactive. It is reoxidized when air is admitted to the illuminated solution. Addition of methyl viologen to the illuminated solution under the exclusion of air yields the blue colour of the semi-reduced methyl viologen, MV^+. As the absorption coefficient of MV^+ is known, the concentration of reduced cadmium can be readily calculated. MV^{2+} reoxidises cadmium atoms:

$$2\,MV^{2+} + Cd^0 \rightarrow Cd^{2+} + 2\,MV^+ \tag{33}$$

The yield of the photocathodic dissolution of CdS in a solution containing 1×10^{-2} M SO_3^{2-} is only 0.005 molecules dissolved per photon absorbed. In the presence of 5×10^{-4} M excess Cd^{2+} ions it amounts to 0.05. Sulfate and dithionate are formed in the ratio 2.2 to 1. The oxidation of SO_3^{2-} is effected by the positive holes produced upon illumination, two holes being necessary to convert one SO_3^{2-} ion into SO_4^{2-} or $1/2\ S_2O_6^{2-}$. If the SO_3^{2-} anion captured the two holes from the same colloidal particle ("two-hole mechanism"), only sulfate would appear as the oxidation product. However, if SO_3^{2-} captured only one hole to form the radical SO_3^-, the final products would be formed by reactions of two such radicals, and these two radicals could even originate from different colloidal particles ("one-hole mechanism")

$$2\,SO_3^- \begin{cases} \nearrow S_2O_6^{2-} \\ \searrow SO_3 + SO_3^{2-} \text{ followed by } SO_3 + H_2O \rightarrow SO_4^{2-} + 2\,H^+ \end{cases} \tag{34}$$

It is known from pulse radiolysis studies that SO_4^{2-} and $S_2O_6^{2-}$ are formed by SO_3^- radicals in homogeneous solution in the ratio 2:1. The fact that practically the same ratio was found in the photo-experiment with CdS sols indicates that the oxidation of sulfite occurs via the one-hole mechanism [75].

The formation of a cadmium atom requires the uptake of two electrons by a Cd^{2+} ion at the surface of a colloidal particle. It seems difficult to believe that Cd^{+} could be the intermediate, as the redox potential of the system Cd^{2+}/Cd^{+} (−1.8 V) is much more negative than the potential of the lower edge of the conduction band in CdS (−0.6 to −0.9 V, depending on pH). In fact, recent pulse radiolysis experiments have shown that Cd^{+} ions are able to transfer an electron to CdS particles (see Sect. 3.7). We therefore propose a mechanism in which a $Cd^{2+}\square$ pair is the intermediate:

$$Cd^{2+}\square + e_1^- \rightarrow Cd^{2+}e_1^- \quad (35)$$

$$Cd^{2+}e_1^- + e_2^- \rightarrow Cd^0 \quad (36)$$

$\square$: anion vacancy; e_1^- and e_2^-: electrons generated in the absorption of a first and second photon. Note that appreciable time may elapse between the generation of e_1^- and e_2^-. The $Cd^{2+}e_1^-$ pair has to survive for this time. In the presence of sulfite, the holes are removed from for the colloid, thus giving $Cd^{2+}e_1$ a chance to survive for quite a while.

Experiments were also performed in which excess electrons were transferred from reducing organic radicals to the colloidal CdS particles. The radicals were formed by γ-irradiation (see footnote on page 117). The amount of cadmium metal was again determined by adding methyl viologen after the irradiation. A high yield of metal was found, practically every two radicals producing one cadmium atom. The higher efficiency of the free radical reaction as compared to the photoreaction is understood in terms of the fact that no positive holes were simultaneously produced in the deposition of electrons on the colloidal particles, i.e. the $Cd^{2+}e_1^-$ pairs were not largely destroyed by recombination with holes as in the photochemical experiment [75].

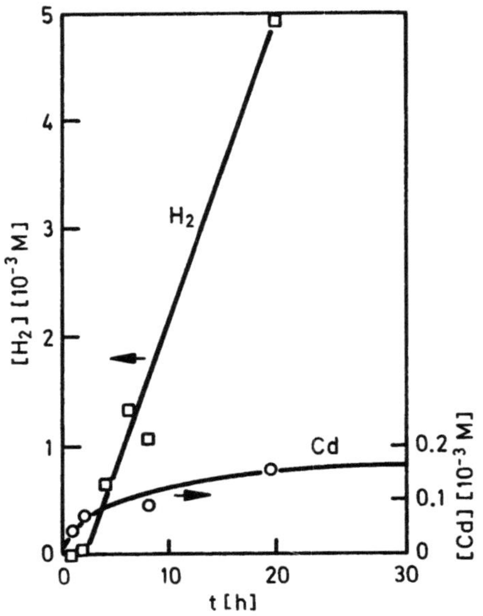

Fig. 12. The hydrogen and cadmium metal concentration at various times of illumination of CdS in a 1 M sulfite solution [75]

The photocathodic dissolution of CdS also occurs in the presence of sodium thiosulfate, but not in the presence of excess SH^- ions. ZnS cannot be dissolved by illumination in the presence of sulfite. However, in the presence of excess Zn^{2+} ions in solution, Zn metal is deposited on the colloidal ZnS particles [61]. This process also occurs when propanol-2 is used as a scavenger for positive holes.

The formation of long-lived $Cd^{2+}e^-$ pairs and of metal atoms which are strongly reducing, offers an explanation for the two-electron reduction of solutes, or to express it in other words, for the intermediate storage of electrons on semiconductor particles. The most interesting two-electron reduction is that of water to form H_2 Eq. (23). Figure 12 shows experimental results which strongly indicate that H_2 formation occurs through intermediate metal formation. A suspension of CdS was illuminated at low light intensity in the presence of 1 M sodium sulfite, the pH of the solution being 11. It is seen that the formation of cadmium metal becomes slower with increasing illumination time. After an induction period during which Cd metal is deposited, the generation of H_2 starts with a quantum yield of 5%. Cadmium sulfide does not have a catalytic surface for hydrogen evolution. It seems that a small amount of deposited cadmium improves the properties of the surface, although cadmium is a metal which has a rather high overpotential for H_2 evolution [75].

3.5 The Problem of Water Splitting and Preparative Aspects

Following the first observations by Darwent and Porter [76,77] numerous studies have been carried out into the generation of hydrogen by illuminating CdS suspensions or CdS sols with visible light [59,60,78–93a]. A sacrificial electron donor such as SH^-, SO_3^{2-}, ethylenediaminetetraacetic acid or cysteine, was used in each of these studies to remove the positive holes from the semiconductor particles. Further, the CdS particles carried a platinum "deposit". This platinization was achieved by stirring the CdS sol with a separately prepared platinum sol, by photochemical reduction of $PtCl_6^{2-}$ in the CdS suspension, or simply by grinding powdered CdS and platinum in an agate mortar. In the presence of the platinum, the yields of H_2 were generally higher by more than a factor of ten than in the solutions that did not contain this co-catalyst. In none of these investigations was explicit proof given for the platinum co-catalyst being tightly attached to the colloidal or suspended particles. It was suggested that the deposited platinum scavenges the electrons produced in the semiconductor particles and catalyses their conversion into hydrogen. The efficiency with which the electrons are scavenged by the platinum deposit was explained by some authors in terms of an electric field-gradient caused by the semiconductor-metal contact [94–96]. However, another mechanism could be operative, in which the sacrificial electron donor efficiently removes positive holes from the illuminated particles, thus leaving long-lived $Cd^{2+}e^-$ pairs or even metal atoms at the surface. The electrons of these pairs or atoms are finally transferred to the platinum particles. This mechanism does not require the platinum to be permanently attached to the semiconductor particles. A similar conclusion has recently been drawn by Bard and coworkers in a study on the H_2 formation on physically mixed CdS/SiO_2 and $Pt/ZnO/SiO_2$ particles in aqueous suspensions containing methanol as hole scavenger [93a].

Cadmium sulfide has been proposed as a catalyst for the splitting of water by sunlight [78,79]. In these studies, the CdS particles were loaded with colloidal RuO_2 in addition to the platinum deposit. RuO_2 is known as an electrode material of low overpotential for the evolution of oxygen from water. Reports about the simultaneous formation of hydrogen and oxygen using the $CdS/Pt/RuO_2$ catalyst could not be confirmed by other laboratories. [56]. In these experiments H_2 was developed and the catalyst seemed to corrode anodically. However, aqueous dispersions of CdS loaded with RuO_2 or Rh_2O_3 evolve oxygen under illumination with visible light in the presence of $PtCl_6^-$ as the electron scavenger [97]. $PtCl_6^-$ is reduced to metallic platinum, which then starts to catalyse the reduction of O_2 by electrons, i.e. the generation of O_2 ceases after a short time of illumination.

Recently an experiment has been reported in which the simultaneous formation of H_2 and O_2 was observed [98]. The sol contained CdS particles loaded with Pt and RuO_2 and μ-peroxo-ruthenium(IV)dioxygen, $[\{Ru(OH)(edta)\}_2(O_2)]$. The latter compound was believed to act as an electron relais in the formation of O_2 from water: a positive hole stored on the RuO_2 deposit oxidizes the relais compound and the resulting oxygen radical of the peroxo-complex oxidizes OH^- to yield O_2. It also was reported that the catalyzing complex did not decompose. However, a closer inspection of the data given in this paper reveals that the decomposition of the peroxo-complex could not have been more than 1 % under the illumination conditions applied. As this seems to be less than could be detected analytically, these experiments also do not give proof that water can be split into H_2 and O_2 using a sensitizer on the basis of CdS.

The generation of H_2 from H_2S by visible light has also been suggested [66,82]. The efficiency of this process is increased by the presence of sulfite. The latter removes the sulfur formed in the oxidation of H_2S to yield thiosulfate. Substitution of RuO_2 by RuS_2 leads to an increased H_2 yield [88]. H_2 from H_2S is also formed in vanadium sulfide dispersions loaded with RuO_2 [88a].

An extraordinary way of stabilizing RuO_2-coated CdS colloids for H_2 generation was chosen by Fendler and co-workers [90,91]. The colloidal particles were generated in situ in surfactant vesicles of dioctadecyldimethylammonium chloride and dihexadecyl phosphate. Thiophenol as a membrane permeable electron donor acted as a sacrificial additive. Later, a surface active re-usable electron donor $(n\text{-}C_{18}H_{37})_2N^+$—$(CH_3)$—$CH_2$—$CH_2$—SH, Br^- was incorporated into the vesicles. Its R—SS—R oxidation product could be chemically reduced by $NaBH_4$ to regenerate the active electron donor. The H_2 yields in these systems were only 0.5 %. However, yields up to 10 % were later reported for a system in which CdS was incorporated into a polymerizable styrene moiety, $(n\text{-}C_{15}H_{31}CO_2(CH_2)_2)\ N^+(CH_3)\ (CH_2C_6H_4CH{=}CH_2)$, Cl^-, and benzyl alcohol was used as the electron donor.

Systems that contain the photocatalyst as a suspended powder or colloid are not convenient in continuous flow or circulation arrangements in which the photoactive system is held in place and can readily be removed for regeneration. CdS can be produced or immobilized in a Nafion membrane together with finely dispersed platinum [99,99a,100]. Hydrogen is produced when the membrane is illuminated in contact with a NaSH solution, the efficiency being comparable to the best colloidal or powdered CdS preparations. The crystal structure of CdS was found to influence greatly the yield with cubic β-CdS acting as a more active catalyst than hexagonal α-CdS.

Meissner et al. [101] proposed a monograin-membrane technique with which short circuits can be avoided at the suspended particles and which makes it possible in principle to separate the anodic and cathodic reactions. The membrane consists of

CdS particles embedded in a polyurethane film. The CdS grains have a diameter of 35–45 μm, whereas the thickness of the polymer membrane is 20 to 25 μm. Each CdS particle protrudes from the membrane surface on both sides and each surface of the membrane can be loaded with Pt or RuO_2. The photoelectrolysis cell consists of two compartments separated by the CdS-monograin membrane. Electrolytic contact between both compartments is provided by an ion-exchange membrane. The expected development of H_2 and O_2 in the two compartments did not take place. Despite this negative result, the principle of compartmenting chemical reactions occurring on small particles using the monograin-membrane technique is worthy of notice for future reference.

While it has not yet been possible to produce O_2 and H_2 simultaneously in the illumination of a colloidal or suspended sulfide, a few reports have appeared about the simultaneous formation of H_2 (making use of conduction band electrons) and interesting organic compounds (making use of the positive holes). Zinc sulfide was used in these experiments, where the electrons are on a more negative potential of -1.4 V and the holes on a more positive potential of $+2.3$ V than in CdS. H_2 is efficiently produced in the presence of both inorganic and organic compounds that react with the positive holes [102–107 b)]. Oxidation of triethylamine yielded 2,3-bis(diethylamino)butane (A) with a small amount of threo-3-(diethylamino)butane-2-ol (B)

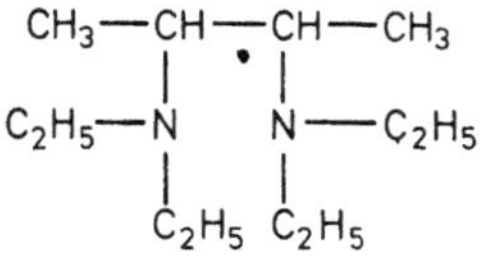

(A)

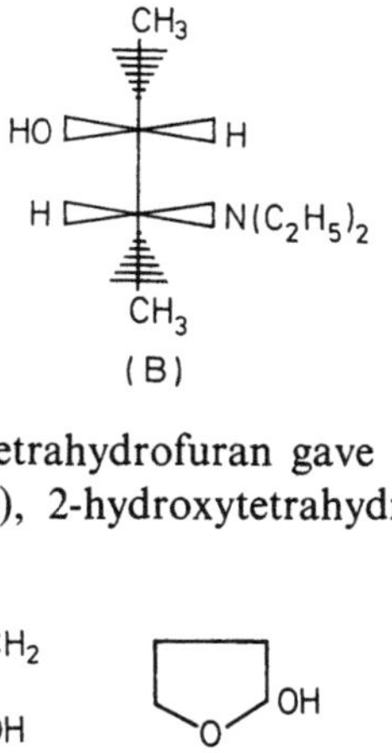

(B)

An equivalent amount of H_2 was also formed. Tetrahydrofuran gave 2,2′-bitetrahydrofuryl (C) with traces of butane-1,4-diol (D), 2-hydroxytetrahydrofuran (E) and unidentified oligomers

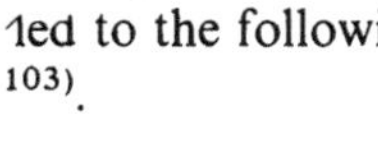

(C)

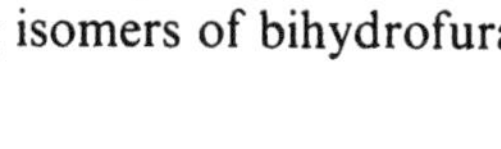

(D)

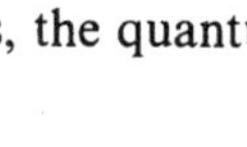

(E)

The alcohols gave the vicinal diols and expected carbonyl compounds. t-Butanol, however, gave only traces of H_2 and the β-carbon-carbon coupling product, 2,5-dimethylhexane-2,5-diol, under comparable conditions [106]. Oxidation of 2,5-dihydrofuran led to the following isomers of bihydrofuranes, the quantum yield being at least 0.1 [103].

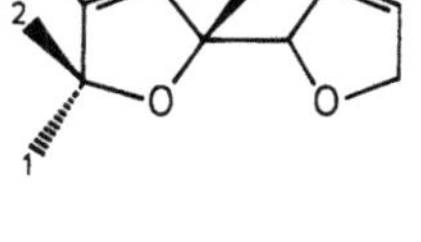

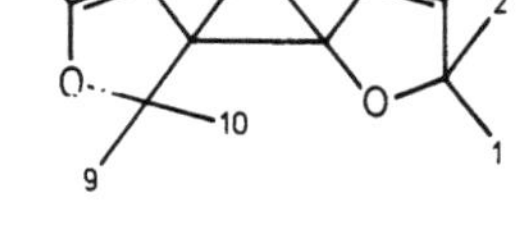

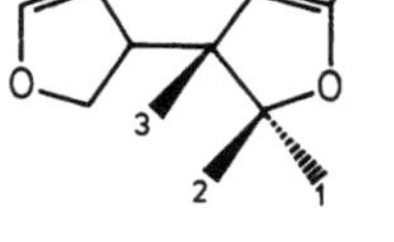

3.6 Some Basic Kinetic Aspects of Interfacial Photoreactions

Certain kinetic aspects of the reactions induced by light absorption in colloidal particles have already been mentioned in the preceding sections, e.g. the storage of charge carriers, and the two-electron and one-electron mechanisms (or two-hole vs. one-hole mechanisms) in the transfer of two charges to a substrate. Further examples are discussed in this section. In addition, we turn our attention to a significant question: When scavengers for positive holes and electrons are present simultaneously in the solution, how is the yield of electron scavenging influenced by the presence of the positive hole scavenger and vice versa? Also, when two scavengers for the same species are present simultaneously, is there competition for the species which can simply be described as in the competition kinetics of homogeneous solutions? In the case of such competition one should be able to obtain relative rate constants for the interfacial reactions on colloidal particles.

We discuss here some reactions on colloidal ZnS particles, which were studied with the aim of finding answers to these questions.

In the illumination of a ZnS sol, containing nitrous oxide and propanol-2, hydrogen and nitrogen are the gaseous products, and acetone and pinacol the condensed products. The acetone to pinacol ratio is 4:1, and the sum of the yields of the condensed products always equals that of the gaseous products. This is explained by the following mechanism [105]:

$$ZnS \xrightarrow{2\,h\nu} ZnS\,(2\,e^- + 2\,h^+) \quad (37)$$

$$2\,e^- \xrightarrow{N_2O + 2H^+} N_2 + H_2O \quad (38)$$

$$2\,e^- \xrightarrow{2\,H_2O} H_2 + 2\,OH^- \quad (39)$$

$$2\,h^+ \xrightarrow{(CH_3)_2CHOH} (CH_3)_2CO + 2\,H^+ \quad (40)$$

$$2\,h^+ \xrightarrow{2(CH_3)_2CHOH} (CH_3)_2CH(OH)CH(OH)CH_3)_2 + 2\,H^+ \quad (41)$$

Figure 13 shows how the yields of the gaseous products depend on the concentrations of the two solutes. In the experiments of the left part of the figure, the N_2O concentration was kept constant at 1.65×10^{-2} M, and the concentration of propanol-2 was varied. The yields of N_2 and H_2 are very small in the absence of propanol-2. With increasing propanol-2 concentration, the N_2 yield increases more rapidly than that of H_2. In the experiments of the right part of Fig. 13, the concentration of propanol-2 was kept constant and that of N_2O varied. With increasing N_2O concentration the H_2 yield decreases while that of N_2 increases, the total gas yield becoming greater. At the highest scavenger concentrations used, i.e. 0.5 M propanol-2 and 1.65×10^{-2} M N_2O, the total gas yield is 0.16 molecules/photon absorbed. As two electrons are required to produce one molecule of H_2 or N_2, the efficiency of electron scavenging is $0.16/0.5 \times 100 = 32\%$.

In the above reaction scheme, reactions 38 and 39 are competing reactions. This competition may formally be described by

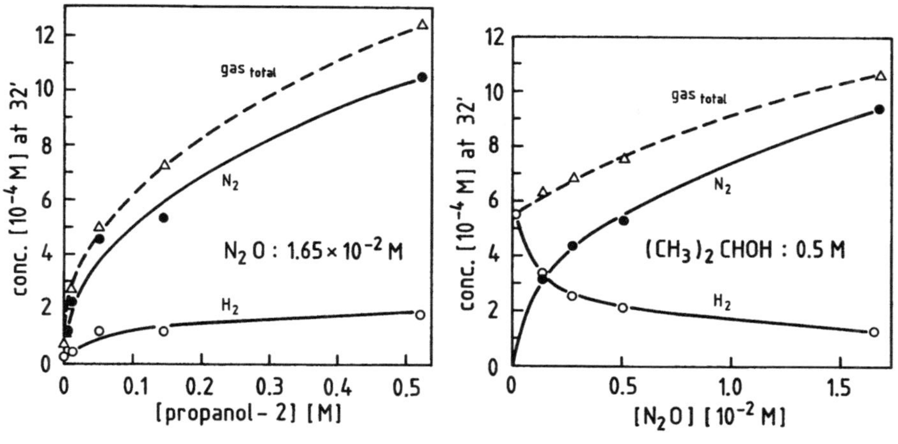

Fig. 13. The concentration of gaseous products at various concentrations of propanol-2 and N_2O in a $2 \cdot 10^{+4}$ M ZnS sol. Illumination time: 32 min [105]

$$[N_2] = [gas_{tot}] \times \frac{k_{38}[N_2O]}{k_{38}[N_2O] + k_{39}} \tag{42}$$

as is usual in the kinetics of competing reactions in homogeneous solution. k_{38} is the rate constant of reaction 38 and k_{39} the pseudo 1st-order rate constant of reaction 39. Rearrangement of Eq. (42) yields the expression:

$$\frac{1}{[N_2]} - \frac{1}{[gas_{tot}]} = \frac{k_{39}}{k_{38}} \frac{1}{[N_2O]\,[gas_{tot}]} \tag{43}$$

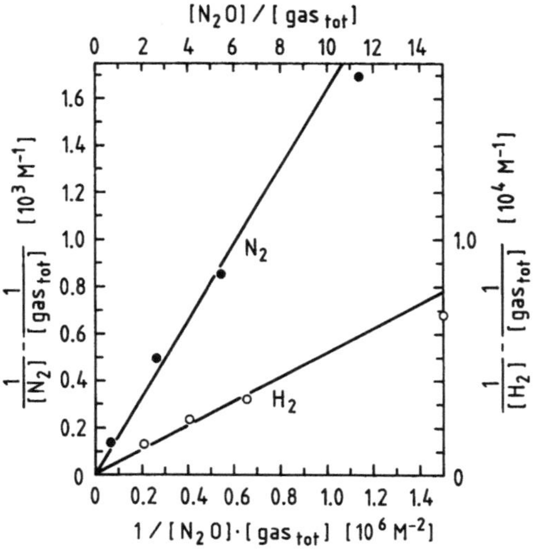

Fig. 14. Plots according to Eqs. (43) and (44) using the data from Fig. 13 [105]

The data of the right part of Fig. 13 were used to obtain the plot in Fig. 14. The ratio k_{39}/k_{38} obtained from the slope of the straight line is 1.65×10^{-3} M. The corresponding treatment gives for the H_2 yield:

$$\frac{1}{[H_2]} - \frac{1}{[gas_{tot}]} = \frac{k_{38}}{k_{39}} \frac{[N_2O]}{[gas_{tot}]} \tag{44}$$

The plot according to Eq. (44) is also shown by Fig. 14. The figure demonstrates that the competition of two scavengers for electrons produced by light absorption in a colloidal particle may be described in a fashion similar to competition in homogeneous solution. Deviations from the straight line behavior in Fig. 14 would be expected in the case where a scavenger is strongly adsorbed at the colloidal particles. Two cases may be distinguished in the scavenging kinetics:

1) Neither the scavenger for electrons nor the scavenger for positive holes are adsorbed at the colloidal particles and both are rather unreactive. Neither of the two can compete efficiently with the recombination of the charge carriers, the result being that the quantum yields of the products are very low.
2) One of the two scavengers is able to interfere with the recombination of the charge carriers, for example by reacting with positive holes. Assume, for example, that the probability p_1 of scavenging the positive hole produced in the absorption of the first photon is 80%. Upon the arrival of the next photon, i.e. the generation of the next $e^- + h^+$ pair, 20% of the colloidal particles will be in the original state, and 80% will carry an excess electron. The probability p_2 of scavenging the holes produced in the absorption of the second photon would be equal to p_1 in 20% of the events, while it will be smaller for the other 80% since the $e^- + h^+$ recombination is expected to be faster in the presence of the excess electron. If the excess electron had no way of escaping from the colloidal particle, the probability p_n after the absorption of the nth photon in this particle, where $n \gg 1$, will be zero. In other words, a photostationary state would be reached in which there existed a constant small concentration of excess electrons on the colloidal particles. One way of escaping is the reduction of water, which is a rather slow process. Again, a stationary concentration of excess electrons will be reached, the probability p_n now being $2 \times \varphi(H_2)$ where $\varphi(H_2)$ is the observed quantum yield expressed as the number of H_2 molecules produced per photon absorbed.

In the system in Fig. 13, propanol-2 is the "main" scavenger which interferes much more strongly with the recombination of the charge carriers than N_2O, since the yields are more drastically dependent on the propanol-2 concentration. In other words, it is mainly the positive hole scavenger that determines the yields, including the yields of the products resulting from the reactions of the electrons with the scavengers H_2O and N_2O. However, the presence of N_2O also influence the yields of the products from hole scavenging. Consider the situation where $[N_2O] = 0$ and only propanol-2 is present (right part of Fig. 13). A constant stationary concentration of electrons exists on the colloidal particles, p_n being very low. With increasing N_2O concentration, the stationary electron concentration becomes smaller and p_n is increased, resulting not only in an increase in $[gas_{tot}]$ but also in the yields of the oxidation products.

It may also be possible that both the e^- and the h^+ scavengers can interfere with the recombination of the charge carriers. Under these conditions, the yields of the products from e^- scavenging are influenced by the presence of the hole scavenger to about the same extent as are the yields of the products of h^+ scavenging, dependent on the concentration of the electron scavenger. Such a system was found in the case of ZnS sols containing propanol-2 and carbon dioxide. CO_2 is reduced to formic acid with a great quantum efficiency (80% in a sol containing 1 M propanol-2 and 0.015 M CO_2), while acetone and pinacol are again formed as oxidation products [105].

The formation of pinacol in ZnS sols containing propanol-2 is indicative of the one-hole mechanism in the oxidation of this solute. In fact, acetone and pinacol are formed in the same ratio as in the oxidation of propanol-2 in homogeneous solution. The first step in the oxidation of propanol-2 must consist of the formation of 1-hydroxy-1-methyl ethyl radicals

$$h^+ + (CH_3)_2CHOH \rightarrow (CH_3)_2COH + H^+ \qquad (45)$$

which subsequently form acetone and pinacol by disproportionation and dimerization. The standard potential of the process $(CH_3)_2CHOH \rightleftarrows (CH_3)_2COH + H^+ + e^-$ is $+1.8$ V. As the upper edge of the valence band in ZnS lies at about $+2.3$ V, reaction 45 is possible from the point of view of energetics. If two holes were picked up by a propanol-2 molecule from the same colloidal particle, only acetone would have been produced. We have thus used the same arguments for proving the one-hole mechanism of the two-hole oxidation of propanol-2 as used in Sect. 3.4 for the oxidation of sulfite Eq. (34).

If CO_2 were reduced via the one-electron mechanism, the intermediate CO_2^- radical would have been formed. Under these circumstances two radicals, i.e. $(CH_3)_2COH$ on the anodic side and CO_2^- on the cathodic side, would have been generated simultaneously on a small ZnS particle, in which case the crossed dimerization product of the radicals should have been formed. It is well known from γ-radiolysis experiments that these products are formed in solutions containing both CO_2 and an alcohol. In the case of propanol-2 as the alcohol, the yield of the crossed combination product is very low (it seems that a CO_2^- and $(CH_3)_2COH$ radical prefer to disproportionate instead of combining). However, when ethanol or methanol are used, the respective crossed combination products, i.e. lactic acid and glycolic acid, are formed in high yields. This can be recognized from Fig. 15 (upper part) where the anion chromatograms of γ-irradiated solutes are shown. The chromatograms show the peaks of the crossed dimerization products besides the peak of the formic acid which is formed in the interaction of two CO_2^- radicals. The lower part of the figure shows the chromatograms for illuminated ZnS sols which contain CO_2 and methanol or ethanol. The crossed dimerization products are absent, formic acid being the only product. One has therefore to conclude that mobile CO_2^- radicals do not appear as intermediates in the photocathodic reduction of CO_2 on illuminated ZnS, or, in other words, CO_2 is reduced via the two-electron mechanism. How this two-electron transfer is brought about is not known in detail. One could assume that a first electron

---

[3] In these solutions, the hydrated electrons from the radiolysis of water produce CO_2^- by their reaction with CO_2, and the OH radicals attack the alcohol to form $(CH_3)_2COH^{\cdot}$ radicals (see also footnote on page 117).

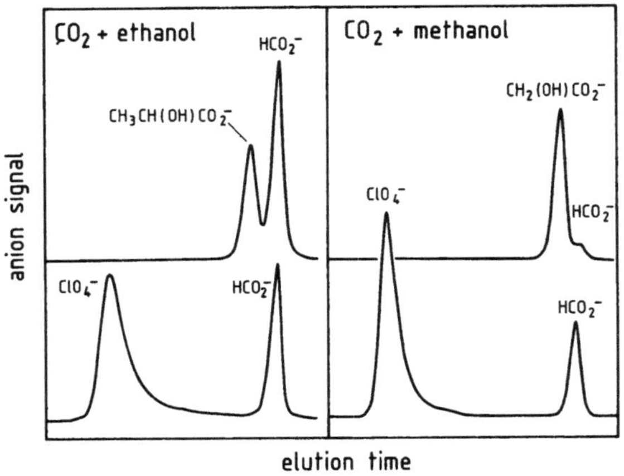

Fig. 15. Anion chromatograms of γ-irradiated solutions of 0.1 M alcohol and 0.015 M CO_2 (upper curves) and of illuminated $2 \cdot 10^{-4}$ M ZnS sols containing 1 M alcohol and 0.015 M CO_2 (lower curve). The latter also contained ClO_4^- anions from the preparation of the ZnS sol [105)]

is scavenged by an adsorbed CO_2 molecule and that the CO_2^- radical formed is so strongly adsorbed that is does not participate in radical-radical reactions and stays on the colloidal particle until the next electron is generated in the absorption of another photon. However, as the redox potential of the system $CO_2^- \rightleftarrows CO_2 + e^-$ is about -2.0 V and the lower edge of the conduction band in ZnS is at about -1.4 V, it hardly seems plausible that CO_2^- can be formed as an intermediate. Another explanation is based on the intermediate formation of Zn^0 [61, 105)]: $Zn^0 + CO_2 + H_2O \rightarrow Zn^{2+} + HCO_2^- + OH^-$.

As a rule, high quantum yields for two-electron transfer reactions are expected when the mechanism is one-electron/two-hole or two-electron/one-hole. In the cases of two-electron/two-hole or one-electron/one-hole efficient back reactions of the intermediates on the colloidal particles or in solution, respectively, will lead to a low yield of the final products.

3.7 Flash Photolysis and Pulse Radiolysis Studies; Non-linear Optical Effects

In laser flash photolysis studies on colloidal semiconductors, irradiation occurs within nanoseconds or even shorter times. A great number of photons may be absorbed in one colloidal particle, the result being a high local concentration of electrons and positive holes, which recombine more rapidly than under continuous illumination conditions. In some colloids the quantum yields depend very strongly on the intensity of the flash. For example, the photoanodic dissolution of Sb_2S_3 Eq. (27) occurs about 100 times less efficiently with intense laser flashes than under continuous illumination. Figures 16 and 17 describe the quantum yield of MV^+ formation in a sol containing methyl viologen, MV^{2+}, and the time profile of the MV^+ radical cation after the flash for various laser intensities. The Sb_2S_3 particles had an agglomeration number of 2000. The overall concentration was 5×10^{-5} M. At the lowest light intensity about one photon was absorbed in one colloidal particle, and the yield for MV^+ formation was almost 30%. At the highest intensity about 50 photons were absorbed, and the

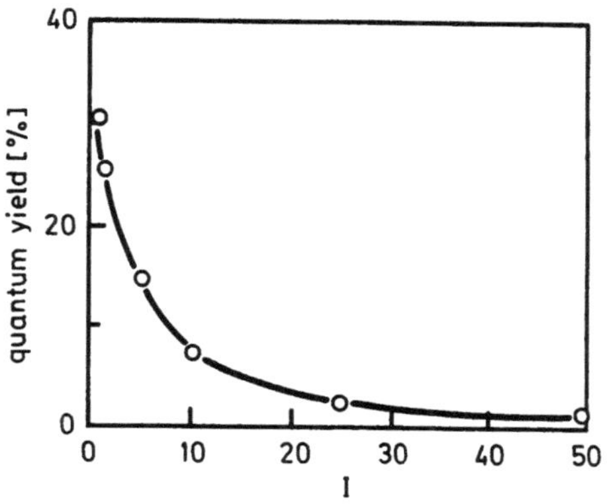

Fig. 16. Quantum yield of MV^+ formation as a function of the laser dose. $5 \cdot 10^{-5}$ M Sb_2S_3 sol, $1 \cdot 10^{-4}$ M MV^{2+}. Dose in photons absorbed per Sb_2S_3 particle [64]

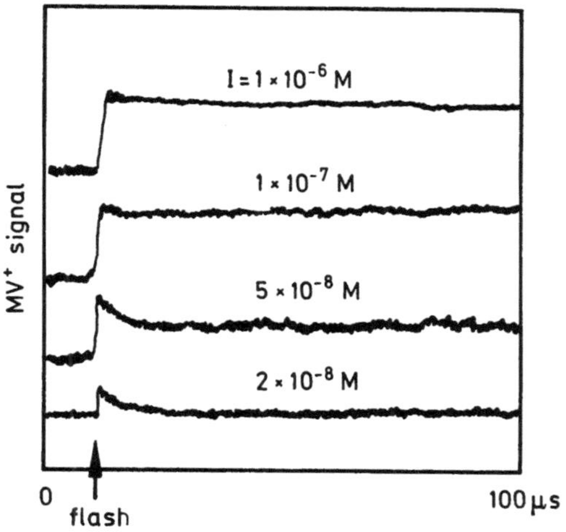

Fig. 17. Time profile of the MV^+ absorption at 610 nm as a function of time for various laser flash intensities (expressed as moles of photons absorbed per liter) [64]

MV^+ yield was only 2.5% (Fig. 16). The MV^+ signal is present immediately after the flash. At low laser intensity the signal decays to practically zero within about 10 μs. The signal decays very little at the highest laser dose (Fig. 17). These observations are explained by the following mechanism [64]. The electrons react during the laser flash with adsorbed MV^{2+} ions: $MV^{2+} + e^- \rightarrow MV^+$. Some of the positive holes oxidize MV^+ ions during the laser flash, and some are trapped. The trapped holes, which are chemically S^- radical anions, react with MV^+ after the flash: $S^- + MV^+ \rightarrow MV^{2+} + S^{2-}$. However, at higher laser intensities, the trapped holes are mainly present as S species resulting from hole-hole interaction: $h^+ + S^{2-} \rightarrow S^-$, followed by $S^- + h^+ \rightarrow S$. The latter species seems to be inefficient in reacting with MV^+.

A surprising observation was made in the first experiments on the flash photolysis of CdS and CdS/ZnS co-colloids [108]. Immediately after the flash from a frequency doubled ruby laser (λ = 347.2 nm; photon energy = 3.57 eV) the absorption spectrum of the hydrated electron was recorded. This species disappeared within 5 to 10 microseconds. More recent studies showed that the quantum yield increased with increasing laser dose. Further, the yield increased with decreasing particle size. The greatest yield observed was 0.07_{aq}^{-} per photon absorbed [110]. These yields are much higher than the quantum yields of 10^{-5} to 10^{-4} observed in photoelectron emission experiments with compact semiconductor electrodes [109]. An electron can be emitted if the energy of the absorbed photon is larger than $E_g + \Delta E$ where E_g is the band gap energy and ΔE the difference in the energies of the hydrated electron and the lower edge of the conduction band. $E_g = 2.4$ eV for CdS; the energy of e_{aq}^- is —2.9 eV on the standard electrochemical scale, and that of the conduction band in CdS is —0.6 to —0.9 eV depending on pH. Thus ΔE is 2.0 to 2.3 eV. The energy of the laser photon was therefore 0.7 to 1.0 eV below the threshold for emission via a one-photons process. e_{aq}^- formation from CdS particles occurs through an absorption process in which two photons are involved.

Grätzel and co-workers found in their first flash experiments that the luminescence of CdS decayed with a lifetime of 0.3 ns [111]. When methyl viologen was present, the signal of the half-reduced electron acceptor, MV^+, was present immediately after the laser flash. With increasing MV^{2+} concentration, the amount of MV^+ formed increased until a plateau value was reached above 5×10^{-4} M MV^{2+}, the overall concentration of the colloid being 1×10^{-3} M (stabilizer: 10^{-2} M hexametaphosphate).

Kuczynski and Thomas [69], who used a sol stabilized with sodium dodecylsulfate, observed that illumination at high intensity levels and high MV^{2+} concentrations leads to two products with absorption maxima at 610 and 540 nm, and which exhibit different kinetics. The 610 nm absorption, which was attributed to MV^+, was present immediately after the flash and decayed thereafter. The 540 nm absorption, which was attributed to the dimer, $(MV^+)_2$, grew rapidly after the flash. The decay of MV^+ was attributed to reoxidation by the residual positive holes. These authors also observed the increase in MV^+ yield with increasing MV^{2+} concentration reported in Grätzel's work. By comparing the increase with the decrease in the luminescence intensity they came to the conclusion that adsorbed MV^{2+} reacts not only with electrons that in the absence of MV^{2+} produce luminescence, but also with electrons that do not give rise to luminescence. Another interesting observation was made on the fluorescence spectrum in the absence of MV^{2+}. While the spectrum had a maximum at 650 nm when excitation occurred in a commercial fluorimeter, the maximum was shifted to 560 nm when excited by the intense laser flash. It was proposed that high incident light intensities tend to saturate the surface states through which the 650 nm emission occurs. More and more charge carriers will then recombine in the bulk, the emitted light quanta having energies close to the band gap energy.

Kuczynski and Thomas [112] also studied the formation of MV^+ in CdS solution using a cationic surfactant, cetyltrimethylammonium bromide, as a stabilizer. In the presence of ethylenediaminetetraacetate (EDTA), the MV^+ yield was greatly increased. EDTA and MV^{2+} form a complex with a resultant negative charge, which is electrostatically bound to the cationic CdS surface. The MV^+ formed is repelled by the positive charge of the cationic stabilizer, a fact which makes the charge separation efficient.

Nosaka and Fox [113] determined the quantum yield for the reduction of methyl viologen adsorbed on colloidal CdS particles as a function of incident light intensity. Electron transfer from CdS to MV^{2+} competes with electron-hole recombination. They derived a bimolecular rate constant of $9 \cdot 10^{-11}$ cm^3 s^{-1} for the latter process.

Serpone et al. studied the fluorescence emitted from a CdS sol containing half-reduced methyl viologen, MV^+, and residual MV^{2+}, the MV^+ being produced by exposing the solution to a conventional light source for 30 min [114]. Two emissions were observed: a fast decaying component with $\tau \approx 0.9$ ns and a slow decaying component with $\tau \approx 7$ ns. The slow emission component sets in at ≈ 1 ns, while the fast emitting transient is formed during the laser pulse. The fast component was attributed to the reaction of a photogenerated hole with adsorbed MV^+, the quantum yield of this emission being 10^{-4} to 10^{-5}: $h^+ + MV^+_{ads} \rightarrow MV^{2+}_{ads} + h\nu$. The slow component was thought to originate from the reaction of photogenerated electrons with cadmium vacancies. These authors also carried out absorption measurements in the sub-nanosecond range. In the presence of MV^{2+} they observed the build-up of the absorption of MV^+ and $(MV^+)_2$ within about 1 ns.

Albery et al. [115] subjected a CdS sol (agglomeration number of particles: 4×10^5) to a 10 µs photo-flash ($\lambda > 300$ nm). Immediately after the flash a negative absorption signal was observed in the wavelength range from 460 to 520 nm, i.e. in the range where CdS begins to absorb. This negative signal increased with the square root of the light intensity. During the long flash of 10 µs, a stationary charge carrier concentration is established as $e^- + h^+$ pairs are produced continuously and disappear via recombination. As the latter process is of second order, the stationary carrier concentration is proportional to the square root of the light intensity. In the absence of air, the signal recovered with a half-life of 50 µs. When 10^{-2} M cysteine, which acts as a scavenger for positive holes, was present, the signal had a recovery time of 10 s. In the presence of O_2, which is an electron scavenger, there was a marked acceleration of the recovery. These experiments show that a small number of long-lived electrons remain on the CdS particles after illumination, an equal number of positive holes having been scavenged either by added solutes or on the surface of the CdS particles. The existence of long-lived electrons on CdS particles under illumination has also been proven by electrochemical experiments. Illuminated suspensions or colloids of CdS containing a supporting electrolyte produce anodic photocurrents at an inert metal electrode immersed in the solution [116–118].

The changes in optical absorption, which the colloidal CdS particles undergo in the presence of excess electrons, were studied in a more quantitative manner by pulse radiolysis [119]. The hydrated electron reacts with CdS particles within some microseconds as can be recognized from the inset of Fig. 18, where the 700 nm absorption of e^-_{aq} is recorded as a function of time. At the wavelength of 470 nm, where the colloidal particles themselves absorb, a decrease in absorption was observed. The bleaching spectrum is shown in Fig. 18. It extends from 500 to 420 nm, i.e. over a relatively narrow wavelength range below the onset of the absorption of CdS. Maximum bleaching at 475 nm occurs with a coefficient of 5×10^4 M^{-1} cm^{-1}. The excess electron obviously influences an optical transition in the colloidal particle as a whole. This was attributed to the influence of the excess electron on the energy of the excitonic state which is formed by light absorption. The energy of this state, which is polarized in the field of the excess electron, is increased, the consequence being a blue-shift in the absorption spectrum. Fig. 19 shows the spectrum of the CdS particles without and with an excess electron. The excess electron could be trapped on a CdS particle in a defect state or chemically be bound to a cation, i.e. in the form of Cd^+. To check the latter possibility, the reaction of Cd^+ ions with CdS particles was investigated. The

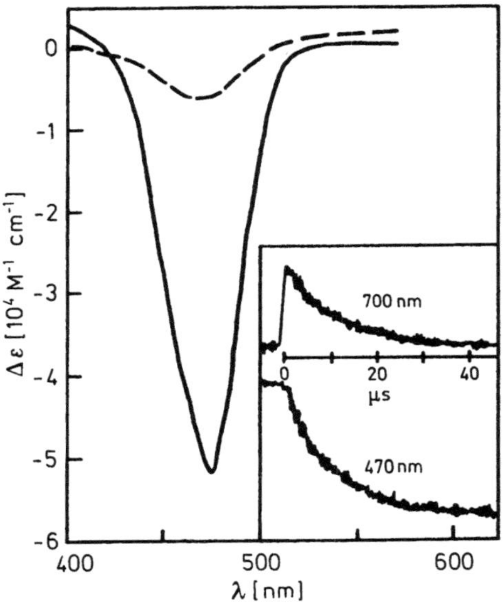

Fig. 18. Bleaching of a CdS sol. Inset: Decay of the 700 nm absorption of e_{aq}^- and time profile of bleaching at 470 nm [119]

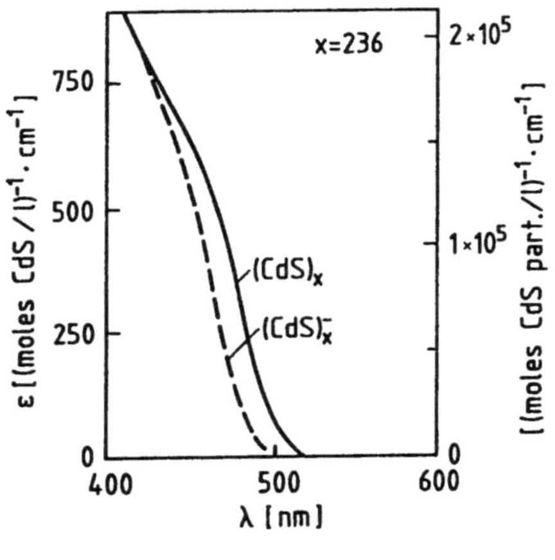

Fig. 19. Spectrum of the absorption edge of CdS particles $(CdS)_x$ and particles carrying one excess electron [119]

Cd^+ ions were formed by pulsing a sol containing some 10^{-4} M excess Cd^{2+} ions. The hydrated electrons then react according to $e_{aq}^- + Cd^{2+} \rightarrow Cd^+$, and the Cd^+ ions which have a strong absorption at 300 nm react with the colloidal particles after the pulse. It was observed that the same bleaching took place during this reaction as in the reaction of e_{aq}^- with CdS particles, and it was concluded from this result that Cd^+ transfers an electron to a CdS particle: $Cd^+ + (CdS)_x \rightarrow Cd^{2+} + (CdS)_x^-$. These observations also are of interest for our understanding of the formation of Cd atoms in the photocathodic dissolution of CdS (see Sect. 3.4). Cd^+ cannot be the intermediate of the overall reaction $2e^- + Cd^{2+} \rightarrow Cd^0$ as already pointed out in discussing the mechanism of Eqs. (35) and (36) [119].

The fact that the absorption spectrum of CdS changes when electrons are produced on it may be regarded as a non-linear optical effect, i.e. a dependence of the absorption coefficient on the light intensity. Other non-linear effects, which are caused by the changes in refractive index with light intensity, are being investigated intensely in semiconductor physics [120]. Such effects also have been found in glasses containing small crystallites of cadmium sulfide-selenide [121] and may be expected to be seen also in colloidal solutions [122, 123].

In a pulse radiolysis study of solutions of CdS of different particle size, the rate of reaction of the OH radical with the colloidal particles was studied as well as the absorption spectrum of the products [124]. The hydroxyl radical reacts on its first encounter with a colloidal CdS particle. Pulse radiolysis can be used to determine the mean particle size if the reaction studied is diffusion controlled [20]. The bimolecular rate constant k_d for such a reaction Eq. (10) is determined by the radius R of the particle and the diffusion coefficient D of the free radical. In the pulse experiment, the half-life $\tau_{1/2}$ of the absorption change accompanying the reaction of the radical with the colloid is measured, and from this half-life an overall bimolecular rate constant $k_2 = \frac{\ln 2}{\tau_{1/2}} \cdot \frac{1}{C_0}$ is obtained, where C_0 is the overall concentration of the colloid. The relation

$$k_2 \cdot n = k_d \tag{46}$$

exists between k_d and k_2, where n is the mean agglomeration number of the colloidal particles. If they are spherical

$$n = \frac{4}{3} \pi L \cdot \frac{\varrho}{M} R^3 \tag{47}$$

where $L = 6.03 \times 10^{23}$ molecules/mole, and ϱ and M are the density and molecular weight of the colloidal material respectively. From Eqs. (10), (46), and (47) one obtains the radius of the colloidal particles in cm as a function of the measured rate constant k_2 (in $M^{-1}\ s^{-1}$):

$$R = \left(\frac{3 \times 10^{-3}\, M \cdot D}{\varrho}\right)^{1/2} \cdot \frac{1}{k_2^{1/2}} \tag{48}$$

Using $D = 2.0 \times 10^{-5}\ cm^2\ s^{-1}$ for the OH radical, the radius according to Eq. (48) was found to agree with the mean radius as determined by electron microscopy [124].

A long-lived product, which has a broad absorption band in the visible, is formed in the attack of CdS by the OH radical. This spectrum is shown by the dashed curve in the upper part of Fig. 20, while curve a in the lower part shows the absorption spectrum of the sol. The CdS particles used in this experiment had a mean diameter of 3.3 nm. The figure also shows spectra for CdS particles of much smaller size where size quantization effects occur. These spectra will be discussed in Sect. 5.4. The broad absorption was attributed to an electron deficient state formed in the attack of OH on the surface of CdS, the chemical nature of this state being the S^- radical anion. In the free state in homogeneous solution, this radical anion absorbs in the UV. However, as a surface species it may absorb visible light by transferring an electron into unoccupied levels of the conduction band of CdS. A similar absorption band was observed in the pulse radiolysis of ZnS and Cd_3P_2 colloids. Note, that bleaching oc-

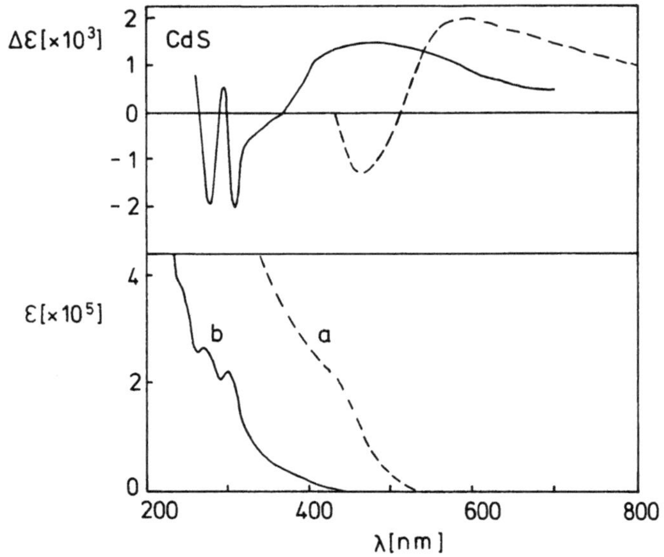

Fig. 20. Absorption spectra of CdS sols of different particle sizes (lower part) and the change in specific absorbance $\Delta\varepsilon$ at 300 μs after the pulse (upper part), mean agglomeration number: (a) = 266, (b) = 94 [124)]

curred in the range from about 450 to 515 nm, i.e. at wavelengths shortly below the onset of absorption of CdS. This means that the onset of absorption of CdS is slightly shifted to shorter wavelengths in the presence of a hole.

In the presence of oxygen, the transient absorption signal (dashed curve in Fig. 20, upper part) decayed, while it was stable for milliseconds in the absence of oxygen. This observation confirmed the conclusion about oxygen being able to react with trapped positive holes in sulfides Eq. (22). As already mentioned in the discussion of the anodic corrosion of colloidal sulfides, oxygen exerts a double action by being an electron scavenger as well as an oxidant of semi-oxidized sulfide anions (Sect. 3.2). The laser flash photolysis of CdS in acetonitrile was studied by Kamat et al [124a)]. Immediately after the flash, a transient bleaching was observed with a maximum at 450 nm. It was attributed to trapped holes as hole scavengers such as triethanolamine or iodide reduced the amount of bleaching and enhanced the rate of bleaching recovery. The absorption of the trapped holes in the long wavelength range of the visible (see above) was also detected.

4 Colloidal Semiconductors: Metal Oxides

4.1 General Remarks

More than half a century ago the first observations were made on photochemical reactions in dispersions of metal oxides. Baur and Perret [125)] found that oxygen was produced and silver deposited in ZnO suspensions illuminated in the presence of silver ions. They interpreted this observation correctly in terms of a molecular electrolysis model, i.e. generation of oxidizing and reducing species in ZnO which subsequently react with water and Ag^+ ions respectively. Similar O_2 evolution was found on WO_3 particles in the presence of Fe^{3+} ions [126]. On the other hand, O_2 is reduced to H_2O_2, and acrylonitrile is photo-polymerized in the presence of ZnO powder [126a]. In more recent times mostly TiO_2 dispersions have been studied. Fujishima and Honda [127] made the discovery that O_2 is produced when a compact TiO_2 electrode is illuminated with near UV light. This discovery triggered a lot of activity in various laboratories to find photoelectrochemical systems that split water into hydrogen and oxygen. In fact, the splitting of water has been shown to occur upon UV irradiation of aqueous suspensions of strontium titanate [128,129]. A great wealth of knowledge about the photocatalytic action of finely dispersed TiO_2 has been accumulated in the pioneering work of Bard and co-workers [130–135]. They have shown that many compounds are decomposed and others synthesized when aqueous or organic suspensions of TiO_2 particles — preferentially platinized at the surface — are exposed to near UV light. Such phototransformations on TiO_2 may even be of interest in preparative and physical organic chemistry [136,137].

The first experiments with colloidal solutions of TiO_2 were reported in the early eighties. First claims [29,138,139,140] that water could be split into H_2 and O_2 using a colloid on which both Pt and RuO_2 were deposited could not be confirmed by other laboratories. The great advantage of using colloidal TiO_2 instead of the powder lies in the possibility of carrying out time-resolved kinetic measurements of interfacial reactions [14,111]. The main purpose of the present article is to review these kinetic and other physico-chemical investigations, while the preparative aspects are described if and when necessary for the sake of completeness.

4.2 Optical Detection of Electrons and Positive Holes in TiO_2

Two kinds of flash photolysis experiments on the interfacial reactions of electrons and holes have been reported. In the first case, the electrons and holes themselves were detected by their optical absorptions, and in the second case the products of their reactions were traced.

Colloidal TiO_2 is made by hydrolysing titanium tetrachloride or titanium tetraisopropoxide. In the author's laboratory a procedure has been developed which allows one to obtain the colloid as a powder from titanium tetraisopropoxide, after careful evaporation of the solvent [141,142]. The powder can be redissolved to give a transparent solution of pH = 3. Alkaline solutions are obtained by the rapid addition of base. In the electron microscope, 5 nm particles can be seen which are often stuck together, the lattice planes in these particles being identifiable. Platinized colloidal TiO_2 can also be made by this method. Colloidal TiO_2 and colloidal Pt sols are mixed, the solvent removed and the powder redissolved. Note that Pt particles do not redissolve when a Pt sol lacking TiO_2 is dried. The hydrolysis of $TiCl_4$ yields large particles (> 100 nm) at room temperature [14] and small particles (5—10 nm) at 0 °C [143,144]. The latter particles can also be recovered as a powder by dialysis of the solution and subsequent vacuum evaporation of the solvent [144a].

Transient signals are not observed when a pure TiO_2 sol is illuminated with a flash from a frequency doubled ruby laser (λ = 347,1 nm). The charge carriers seem to

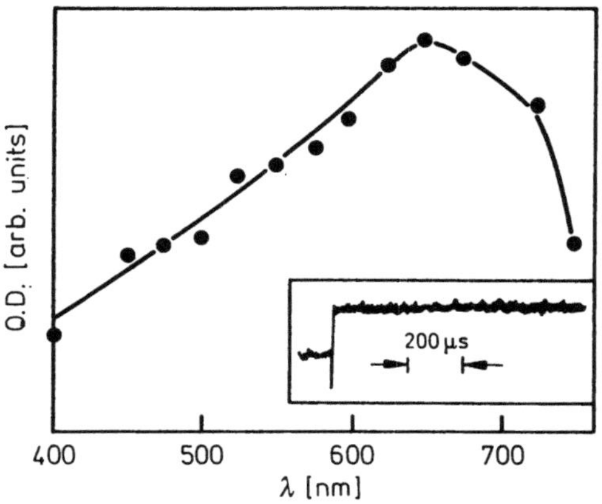

Fig. 21. Absorption spectrum of a laser-flashed solution of 6.3×10^{-3} mol dm^{-3} TiO_2 containing 5×10^{-3} mol dm^{-3} PVA and the time profile of the absorption signal (inset) at pH 10 [141]

recombine rapidly during the flash of 15 ns. However, when the solution contains polyvinyl alcohol a long-lived absorption signal is present immediately after the flash [141, 142]. Figure 21 shows the absorption spectrum, which was attributed to a reducing intermediate of the photolysis of TiO_2. In the presence of O_2 or other electron acceptors the absorption decays after the flash. The broad absorption band with a maximum at 650 nm corresponds to a blue colored solution. Such a color has indeed been seen in γ-irradiation experiments on electron transfer from reducing organic radicals to colloidal TiO_2 particles [14] and in prolonged illumination of TiO_2 sols containing polyvinyl alcohol [111]. The position of the maximum of the absorption band was red-shifted with increasing size of the TiO_2 particles. Illumination of the blue TiO_2 sol with visible light ($\lambda > 500$ nm) led to bleaching. Addition of tetranitromethane to the blue sol produced the yellow color of $C(NO_2)_3^-$, a well-known reduction product of tetranitromethane, and from such experiments it was concluded that two kinds of reducing species were present [14]. It has not yet been clearly decided whether the species are free or trapped electrons or both. The absorption of free electrons should lie mainly in the infrared. It was also supposed that the blue color arose from hydrated Ti_2O_3 [14]. Ti^{3+} ions have indeed been observed in a more recent ESR study of illuminated TiO_2 sol [145]. Similar observations have been made with a WO_3 colloid (Sect. 4.6).

The polyvinyl alcohol in the experiment of Fig. 21 plays the role of the "main" scavenger in the terminology of Sect. 3.6. It is strongly adsorbed on the colloidal particle and able to interfere with the recombination of the charge carriers by scavenging positive holes. Excess electrons can therefore remain on the colloidal particles for a long time. The main scavenger reacting with the holes determines the yield of the reaction of other additives with the electrons.

Another intermediate of the photolysis of TiO_2 was observed in experiments with platinized particles (in the absence of polyvinyl alcohol). The spectrum shown in Fig. 22 is present immediately after the laser flash. The signal decays as shown by the inset in the figure. The rate of decay is not influenced by oxygen but is increased by oxidizable compounds such as Br^- ions in the solution. The broad absorption band in Fig. 22 with a maximum at 430 nm was attributed to trapped positive holes. Chemically, a trapped hole is an O^- radical anion. In homogeneous aqueous solution, O^-

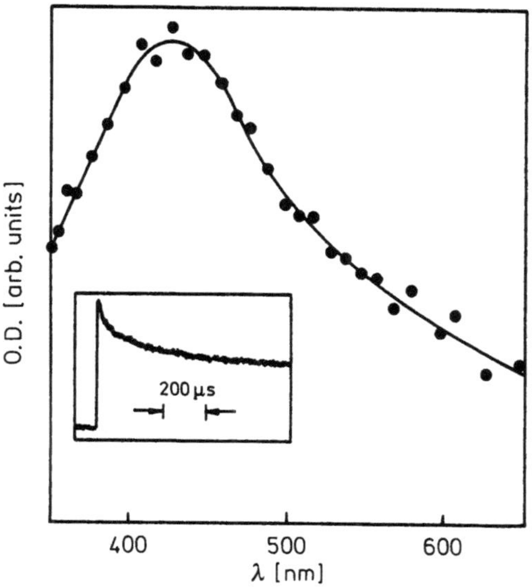

Fig. 22. Absorption spectrum of a laser-flashed TiO_2—Pt colloid (3.8×10^{-3} mol dm^{-3} TiO_2 and 1.6×10^{-5} mol dm^{-3} Pt) and the decay of the absorption (inset) at pH 2.5 [141]

absorbs in the UV. However, when it is part of a TiO_2 particle it may absorb in the visible. The absorption could be caused by the transition of an electron from the valence band to the O^- radical or by an electron transfer from O^- into unoccupied levels of the conduction band. The spectrum of the hole can also be observed in alkaline solutions using pure TiO_2 with methyl viologen, MV^{2+}, as additive. Above pH = 6 the TiO_2 particles carry a negative excess charge and MV^{2+} ions are strongly adsorbed. Immediately after the flash the intense absorption of MV^+ is seen. However, when oxygen is present in the solution, MV^+ is rapidly reoxidized and the spectrum of the hole can be recognized. Methyl viologen in alkaline solution and platinum in acidic solution play the role of the "main" scavenger by reacting with the electrons formed during the laser flash. Excess positive holes can then reside on the colloidal particles for quite a while. The decay of the absorption of the holes is ascribed to their reaction with water. In acidic solutions the holes live for milliseconds, in alkaline solution only for microseconds [141, 142]:

$$h^+ + OH^- \rightarrow OH \tag{49}$$

The reaction of the trapped holes with Br^- ions adsorbed at the colloidal particles was used to determine the absorption coefficient of the holes. In the presence of Br^- less absorption of the holes is observed after the flash, as the Br^- ions react with some of the holes during the laser flash. On the other hand, the absorption of Br_2^-, i.e. the oxidation product of Br^-, can be seen. By comparing the decrease in h^+ absorption after the flash with the absorption of Br_2^-, and knowing the absorption coefficient of Br_2^-, the absorption coefficient of the hole, $\epsilon_{475\,nm} = 9.2 \times 10^3\ M^{-1}\ cm^{-1}$, was calculated. Electron deficient surface states are well known from studies in electrochemistry where their existence is deduced from electric current measurements. Studies on colloids complement these findings by characterizing the trapped positive holes through their optical absorption. The trapped holes can now be dealt with in flash photolysis experiments, as are the intermediates of photoreactions in homogeneous solution.

The oxidation of various organic additives to the platinized TiO_2 sol was also studied by recording the absorption of the holes after the laser flash. Some of the holes react with adsorbed molecules during the flash. This reaction is governed by the ability of the organic compound to be adsorbed on the TiO_2 particles and not by the oxidizability of the compound which is known from oxidation experiments in homogeneous solution. The potential of the free positive holes is as high as 3.1 V. These holes react with adsorbed organic compounds rapidly, regardless of the oxidation potential of the substrate, as there is always a great driving force available. The holes that are trapped react after the flash and the rate of this reaction depends on the specific oxidizability of the substrate. Trapped holes have less positive potential, therefore less driving force is available for the redox reaction, and the rate of reaction is dependent on the redox potential of the substrate. Ethanol, for example, is only slightly adsorbed at the colloidal particles and therefore cannot efficiently scavenge the holes before they are trapped, so that it reacts with the trapped holes after the flash. Citrate, on the other hand, is strongly adsorbed and reacts with the holes mainly during the flash [142].

Grätzel and Serpone and co-workers [146] recently reported on a picosecond laser flash photolysis study of TiO_2. They observed the absorption spectrum immediately after the 30 ps flash and attributed it to electrons trapped on Ti^{4+} ions at the surface of the colloidal particles. The absorption decayed within nanoseconds, the rate being faster as the number of photons absorbed per colloidal particle increased. This decay was attributed to the recombination of the trapped electrons with holes.

4.3 TiO_2 Particles and the MV^{2+}/MV^{+} System

The factors that control interfacial electron transfer on colloidal TiO_2 particles have been studied by several authors who used methyl viologen, MV^{2+}, or semi-reduced methyl viologen, MV^{+}, as the reactant. The standard redox potential of the MV^{2+}/MV^{+} system is -0.44 V, and the potential of the conduction band in macrocrystalline TiO_2 is 0.06 V at $pH = 0$. With increasing pH, the potential in TiO_2 is known to become more negative by 59 mV per pH unit. The electron transfer from illuminated TiO_2 particles to MV^{2+} in solution was studied by flash photolysis, and the transfer from MV^{+} to TiO_2 by pulse radiolysis. In acidic solutions where the TiO_2 particles are positively charged, MV^{2+} is not adsorbed and the electron transfer therefore requires the diffusion of MV^{2+} to the illuminated colloidal particles.

Grätzel and co-workers [111] found that the 600 nm absorption of MV^{+} is built up after the laser flash. The colloid was stabilized by polyvinyl alcohol. The laser flash produced a large number of electrons in each colloidal particle. The build-up followed a first order rate law, the rate constant being proportional to the MV^{2+} concentration, and the final amount of MV^{+} formed also increased with the MV^{2+} concentration. Figure 23 shows the final MV^{+} concentration as a function of the pH of the solution. Below $pH = 2$, MV^{2+} is not reduced. The electron transfer from the colloidal particles proceeds until electrochemical equilibrium is reached between the TiO_2 particles and the MV^{2+}/MV^{+} system in solution. At low pH values this equilib-

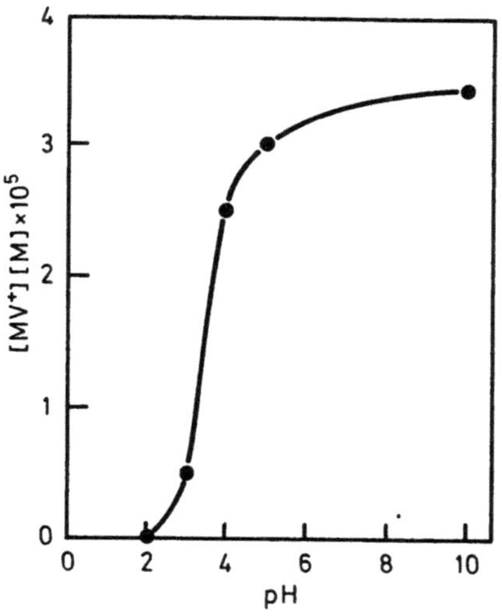

Fig. 23. Effect of pH on the yield of MV^+ after completion of interfacial electron transfer. Colloidal TiO_2 (500 mg/L); $[MV^{2+}] = 10^{-3}$ M [111]

rium is displaced towards the electrons on the colloidal particles due to the anodic shift of the potential of the TiO_2 particles with increasing proton concentration. Under these conditions only a small fraction of the conduction band electrons produced by laser excitation will leave the TiO_2 particle and form MV^+. From the observed data, a potential of the conduction band of TiO_2 particles of —0.14 V (pH = 0) was derived.

The rate of MV^+ formation was also dependent on pH. The bimolecular rate constant, as calculated from the first order rate constant of the MV^+ build-up and the concentration of colloidal particles, was substantially smaller than expected for a diffusion controlled reaction Eq. (10). The electrochemical rate constant k′ Eq. (9) which largely determines the rate of reaction was calculated using a diffusion coefficient of MV^{2+} of 10^{-5} cm^2 s^{-1} [147]. A plot of log k′ vs. pH is shown in Fig. 24. The usual Tafel evaluation yielded a transfer coefficient $\alpha = 0.52$ and a rate constant k'_0 of 4×10^{-3} cm · s^{-1} at the standard potential of the MV^{2+}/MV^+ couple. This k'_0 value corresponds to a moderately fast electrochemical reaction. In this electrode-kinetic treatment the changes in the rate of electron transfer with pH were attributed only to the changes in the overpotential. A more exact treatment should also take into account the electrostatic effect on the rate of reaction which also changes with pH.

The kinetics are quite different when an MV^{2+} derivative is used that carries a long hydrocarbon chain [148]. This electron acceptor adheres strongly to the surface of the TiO_2 particles and the electron transfer occurs very rapidly. As the half reduced MV^+ derivative is also adsorbed, a second electron can be picked up to form the MV^0 derivative.

In the pulse radiolysis studies on the reaction of MV^+ with TiO_2, the sol contained propanol-2 or formate and methyl viologen, MV^{2+} [149]. Ionizing radiation produces reducing organic radicals, i.e. $(CH_3)_2COH$ or CO_2^-, respectively, and these radicals react rapidly with MV^{2+} to form MV^+. The reaction of MV^+ with the colloidal particles was then followed by recording the 600 nm absorption of MV^+. The rate of reaction was found to be slower than predicted for a diffusion controlled reaction.

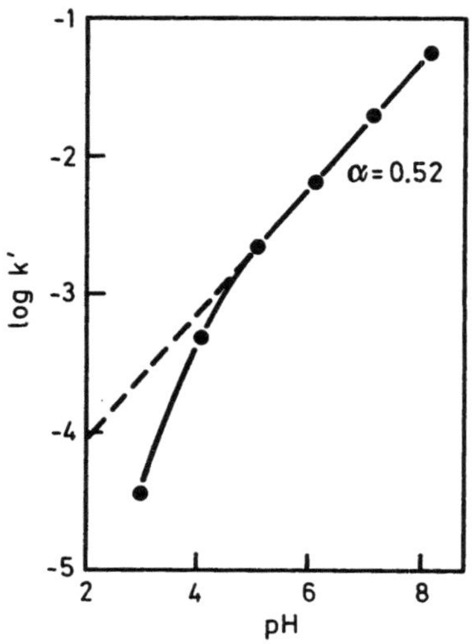

Fig. 24. Plot of the specific rate of interfacial electron transfer k' vs. the pH of the solution. Experiments were conducted with colloidal solutions of TiO_2 (R = 10 nm) in the presence of 10^{-3} M MV^{2+} [111]

With increasing pH the reaction became less complete as the reverse reaction $MV^{2+} + TiO_2^{x-} \rightarrow MV^{+} + TiO_2^{(x-1)-}$ became more important. A potential of the conduction band of the TiO_2 particles of —0.1 to —0.2 V (pH = 0) was derived from these measurements. Stabilization of the particles by polyvinyl alcohol did not change this potential. Particles of 80 nm diameter had the same potential as particles of 7 nm.

Darwent and co-workers [150,150a] have drawn attention to the fact that the surface area of a colloid will have a significant effect on the rate of the interfacial electron transfer reactions, which are governed by the transfer itself and not by diffusion. Equation (9) simplifies to give

$$k_2 = 4\pi R^2 L k' \qquad (50)$$

for small values of k' (with $k_1 = 0$). The colloids always contain a discrete distribution of particle radii. There will be a distribution of rate constants, i.e. simple monoexponential kinetics should not be expected for the observed reactions. Based on a general model for dispersed kinetics in heterogeneous systems [151], a formula was derived that describes the kinetics in terms of two parameters, i.e. the number-average radius in a Gaussian distribution and the spread of the Gaussian [150]. Deviations from the monoexponential kinetic curves in the work of various authors could be explained in this way. Darwent et al. [150,152] also investigated how pH variations affect the electrostatic charge that is experienced as an ionic reagent approaches a colloidal particle. As already mentioned above in the discussion of Grätzel's results, changes in pH will influence the rate of reaction of charged electron acceptors, and k' has to be corrected for this electrostatic effect. The transfer coefficient α can be expressed as

$$\alpha = \alpha_0 + z\beta \cdot I^{-1/2} \qquad (51)$$

where α_0 is the value when there is no electrostatic effect (infinite ionic strength I), β is a constant that depends on the rate of change of TiO_2 surface charge with pH,

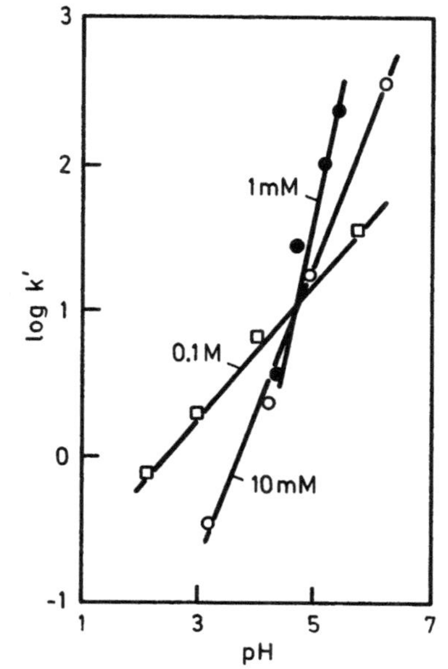

Fig. 25. Variation in pseudo-first-order rate constant (k') for reduction of MV^{2+} with pH [150]. Ionic strength values are given on each line

and z is the charge on the redox agent (+2 for MV^{2+}). Fig. 25 shows how the averaged first order rate constant of the reaction of illuminated TiO_2 with MV^{2+} changes with pH at different ionic strengths. The point of zero charge of the colloid was at pH = 4.5. All the curves cross at this pH. Measurement of the rate of electron transfer under various pH and ionic strength conditions may be used as a method for determining the point of zero charge of a colloid.

4.4 Various Flash Photolysis Studies on TiO_2 Sols

The halide anions Cl^-, Br^-, and I^- are oxidized on illuminated Pt—TiO_2 powder [153, 154]. Flash photolysis showed that Cl_2^-, Br_2^- and I_2^- are the intermediates which are readily traced by their characteristic absorption spectra [14, 143]. Similarly, SCN^- is oxidized to $(SCN)_2^-$. The absorptions of these species were present immediately after a laser flash, which indicates that adsorbed ions reacted with the holes. After the flash the absorptions decayed according to a second order rate law as the final products of oxidation were formed, for example: $2\,Br_2^- \rightarrow 2\,Br^- + Br_2$. Figure 26 shows the yield as a function of the concentration of the sodium halides in the TiO_2 sol. The yields increase and approach a limiting value at higher concentrations. The shape of these curves is determined by the adsorption isotherm of the anions on TiO_2. With increasing pH, the yield rapidly decreased which is explained by a lower degree of adsorption owing to the decreased positive charge of the colloidal particles. In the experiments of Fig. 26a colloid of large particle size (> 100nm) was used [14]. Greater quantum yields of oxidation were observed for a colloid of smaller particle size [143].

Tetranitromethane is reduced to the colored stable anion $C(NO_2)_3^-$ when a TiO_2 sol containing tetranitromethane, $C(NO_4)_4$, and SCN^- is flashed. While the absorp-

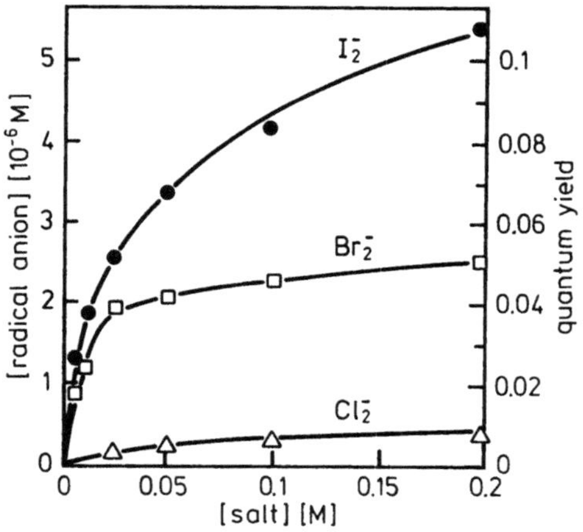

Fig. 26. Concentration of radical anions immediately after the laser flash and quantum yield as functions of the concentration of the respective sodium halide [14)]

tion of $(SCN)_2^-$ appears immediately after the flash, that of $C(NO_2)_3^-$ is built up afterwards. In this system the main scavenger is SCN^- which is adsorbed and interferes with the recombination of the charge carriers. The excess electrons react with tetranitromethane after the flash [142)].

The holes of TiO_2 are able to oxidize carbonate to form CO_3^- radicals which can be analysed by their characteristic absorption at 600 nm. Under steady-state illumination formaldehyde is produced [155)].

Organic dyes such as thiazine and oxazine are reduced upon band gap excitation of a TiO_2 sol in water or acetonitrile, added SCN^- increasing the yield. The first product of reduction is the semi-reduced radical ion, with quantum yields up to 0.1. Further disproportionation leads to the stable leuco dye. Continuous illumination of TiO_2 particles generates the leuco dye in an amount sufficient to produce significant changes in the electrochemical potential of the solution. Together with a non-illuminated dye solution a photoelectrochemical cell may be constructed [156–158)]. A detailed study was also made on the reduction of cobaltoceniumdicarboxylate on colloidal TiO_2 particles (pH = 10) which were stabilized by polyvinyl alcohol [159)]. Reaction occurred after the laser flash, polyvinyl alcohol acting as the main scavenger. Very efficient electron transfer was observed in experiments with a polymer that carried a great number of methyl viologen groups. Half reduced methyl viologen was formed during the flash with a quantum yield close to one [160)]. Both cobaltoceniumdicarboxylate and methyl viologen have been proposed as electron relays in devices for solar energy conversion [161)].

Flash photolysis with microwave detection of charge carriers could become an additional technique in the future. The method has not yet been applied to colloids but has been used with small suspended particles. Immediately after the laser flash a conductivity signal was observed which decayed in the 0.1 to 1 microsecond range. The signal was longer-lived for a suspension of TiO_2 in para-dioxane than in Decalin. Such an effect of the surrounding medium on the decay kinetics of the conductivity indicates that surface states are involved [162)].

Transient Raman spectroscopy was also used to study charge transfer reactions across aqueous solution interfaces. One optical pulse above the band gap creates electrons and holes which react with adsorbed substrate molecules. After a fixed delay of about 5 ns, a second pulse below the band gap generates Raman spectra of the surface reaction products. One-hole transfer to SCN^- on TiO_2 was observed as well as electron transfer to MV^{2+} on TiO_2 and CdS and to $C_{14}MV^{2+}$ on TiO_2. The Raman spectra were indistinguishable from those of the same species fully solvated in water, and no evidence for surface-enhanced Raman scattering was found. The authors concluded that electrons tunnel from the conduction band to the adsorbed ions which retain their first solvation layer in the adsorbed state [163, 164].

Organic dyes often are strongly adsorbed on colloidal TiO_2 particles. They can induce charge transfer into the conduction band of TiO_2 upon absorption of light. Processes of this kind were studied using erythrosine [165], eosin [166, 166a], rosa bengal [167], chlorophyllin [168] and RuL_2^{2+} (L = diisopropyl-2,2′-bipyridine-4,4′-dicarboxilate) [169]. Flash photolysis was the method applied in most of these studies, although resonance Raman spectroscopy [170] and microwave measurements [167] were also carried out. The product of charge injection is an electron in the conduction band of the colloid and an adsorbed cation of the dye. The charge injection from the singlet state of the dye generally is more efficient than from the triplet state. In the case of efficient charge injection, the fluorescence of the dye is quenched. At high occupancy of a TiO_2 particle by the dye, the quantum yield of charge injection may be decreased, this effect arising from concentration quenching. The electron-cation radical pairs formed recombine on the TiO_2 particle where they were formed unless the cation is rapidly desorbed. In the latter case the lifetime of the separated charge carriers is strongly increased.

4.5 Hydrogen and Oxygen Generation on TiO_2

Hydrogen is readily formed in the near UV illumination of sols or suspensions of platinized TiO_2 containing a sacrificial electron donor [171–176]. Figure 27 shows the H_2 concentration as a function of time for various alcohols as electron donor. In the presence of primary and secondary alcohols the rates of H_2 generation are equal although the induction periods at the start of illumination are different. In the presence of a tertiary alcohol the rate of H_2 formation is substantially lower. As mentioned in section 4.2 alcohols are not strongly adsorbed at the TiO_2 particles and therefore do not react with the positive holes before they are trapped. The main scavenger in this system is the platinum deposit which scavenges electrons and catalyses the reduction of hydrogen ions: $2\,e^- + 2\,H^+ \xrightarrow{Pt} H_2$. A surprising observation was that the H_2 yield decreased with decreasing pH (in the range 3.0 to 1.5), i.e. the efficiency of recombination of the charge carriers was increased with increasing hydrogen ion concentration. This could be due to less efficient hole trapping at the surface and a longer residence time of the electrons in the TiO_2 particle before they reach the Pt deposit.

Numerous papers have dealt with the problem of simultaneous production of hydrogen and oxygen on illuminated colloidal or suspended TiO_2 particles. It is not intended to review this problem in a comprehensive manner, but a few important details may be mentioned. H_2 and O_2 were reported to be generated in the band gap illumination of platinized powdered TiO_2 in contact with water, although the mecha-

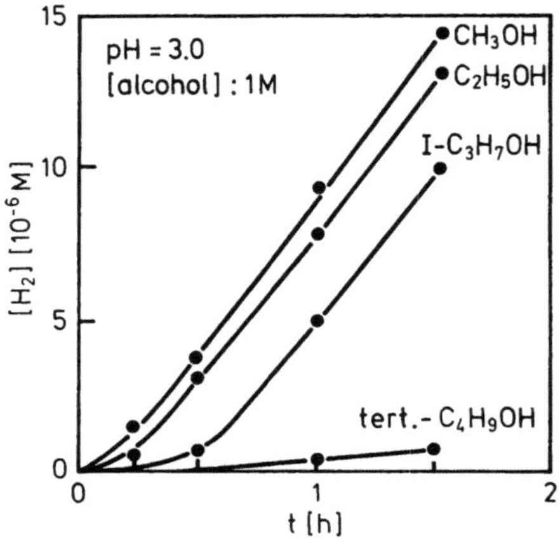

Fig. 27. Hydrogen formation in a TiO_2 sol containing various alcohols [142]

nism is not well-understood [177]. In many investigations a RuO_2 deposit on the TiO_2 particles in addition to the Pt deposit was used in order to facilitate the anodic process: $4\,h^+ + 2\,H_2O \xrightarrow{RuO_2} 4\,H^+ + O_2$ [29,178-181]. RuO_2 is known as an electrode material of low overpotential for O_2 evolution. The cleavage of water by visible light was also attempted by adding ruthenium tris-bipyridylium chloride, $Ru(bpy)_3^{2+}$, as a light absorbing dye and methyl viologen, MV^{2+}, as an electron relay to the TiO_2—Pt—RuO_2 sol [29]:

$$Ru(bpy)_3^{2+} + h\nu \longrightarrow Ru(bpy)_3^{2+*} \tag{52}$$

$$Ru(bpy)_3^{2+*} + MV^{2+} \longrightarrow Ru(bpy)_3^{3+} + MV^+ \tag{53}$$

$$MV^+ + H^+ \xrightarrow{Pt-TiO_2} MV^{2+} + 1/2\,H_2 \tag{54}$$

$$Ru(bpy)_3^{3+} + 1/2\,H_2O \xrightarrow{RuO_2} H^+ + 1/4\,O_2 + Ru(bpy)_3^{2+} \tag{55}$$

In fact, the formation of H_2 was observed, and it was assumed that the stoichiometric amount of O_2 must have been generated too. However, the oxygen was not measured and the question then arose whether water was indeed decomposed into $H_2 + O_2$, the O_2 escaping analysis, or whether an unknown sacrificial electron donor was present which interfered with reaction 55.

With respect to using methyl viologen as electron relay, it might be of interest to note that MV^{2+} can be oxidized by positive holes produced in illuminated colloidal semiconductors such as TiO_2 [142]. Two oxidation products of MV^{2+} are 1′,2′-dihydro-1,1′-dimethyl-2′-oxo-4,4′-bipyridylium chloride and 3,4-dihydro-1,1′-dimethyl-3-oxo-4,4′-bipyridylium chloride, which can readily be detected by their strong fluorescences at 516 nm and 528 nm, respectively. These products are also produced in the direct photolysis of MV^{2+} solutions [182] and in the reaction of MV^{2+} with OH radicals in homogeneous solution [183].

In more recent papers, it has been proposed that water is not decomposed into $H_2 + O_2$ on TiO_2 particles, but into $H_2 + H_2O_2$ [184-188]. Hydrogen peroxide is known to be specifically adsorbed at the surface of TiO_2 [189]. When TiO_2 particles are illuminated in the presence of oxygen, O_2 is consumed, but H_2O_2 cannot be detected in the solution using catalase [190]. Using redox indicators such as permanganate or o-dianisidine, it could be shown that peroxo species were formed during the photo-uptake of O_2 and peroxide is produced concomitantly with hydrogen during photolysis of TiO_2—Pt dispersions [185-187]. In the presence of Ba^{2+} ions, which form an insoluble peroxide, an increased H_2 yield was observed [188].

4.6 Other Metal Oxides: Fe_2O_3, WO_3, ZnO, CdO, In_2O_3, MnO_2

Colloids of α-Fe_2O_3 are made by hydrolysis of $FeCl_3$ and subsequent dialysis of the sol. Polyvinyl alcohol is often used as a stabilizing agent. The band gap in Fe_2O_3 is 2.2 eV. In some of the studies on colloidal Fe_2O_3 free radicals were generated by ionizing radiation and electron transfer reactions with the colloidal particles investigated. Buxton et al. [191,192] observed a cathodic dissolution of α-Fe_2O_3 in acidic solution in which $(CH_3)_2COH$ radicals were produced (see footnote on page 117):

$$(CH_3)_2COH + Fe^{3+}_{coll} \rightarrow Fe^{2+}_{aq} + (CH_3)_2CO + H^+ \quad (56)$$

A similar study was conducted by Dimitrijević et al. [149] with neutral solutions of α-Fe_2O_3. The yield of Fe^{2+}_{aq} was found to be very low. However, a large Fe^{2+} yield was found after dissolution of the colloid by hydrochloric acid under an argon atmosphere. This showed that electrons donated by the free radicals penetrated deep into the colloidal particles to reduce iron to Fe^{2+}_{coll}. Buxton et al. [192] observed in a study on the reductive dissolution of colloidal Fe_3O_4 that Fe^{2+} ions in this material are less readily released into the aqueous phase than reduced Fe^{3+} ions.

Pulse radiolysis studies showed [149] that the rate of the reaction of MV^+ with α-Fe_2O_3, in which an electron is transferred to the colloidal particles, is slower than predicted for a diffusion controlled reaction. For pH > 8, the reaction is incomplete as the reverse reaction $Fe_2O_3^- + MV^{2+} \rightarrow Fe_2O_3 + MV^+$ takes place more efficiently and shifts the equilibrium between electrons on Fe_2O_3 and the MV^{2+}/MV^+ redox couple to the MV^+ side. As in the case of the corresponding studies on the TiO_2-MV^+ system (Sect. 4.3) this effect was explained by the cathodic shift in the flat band potential of electrons in Fe_2O_3 with increasing pH. It does not matter in these equilibrium considerations whether the injected electrons are free or trapped. This is because equilibrium is established between the semiconductor particle and the redox couple in solution. Under these conditions only the position of the Fermi level in the semiconductor that is equilibrated with the redox couple is important. The flat band potential of colloidal α-Fe_2O_3 was found to lie at —0.1 to —0.2 V.

Moser and Grätzel [143] reported that the absorption signals of I_2^- could be seen immediately after a laser flash on an I^--containing α-Fe_2O_3 sol. The quantum yield was 80%. In a subsequent reaction, I_2^- disappeared according to a second order rate law: $2\,I_2^- \rightarrow 2\,I^- + I_2$. The authors pointed out that single crystals and polycrystalline α-Fe_2O_3 electrodes show only small efficiencies as photoanodes for the oxidation of water and other substrates. This arises from the low mobility of the charge carriers and a short hole diffusion length. In the α-Fe_2O_3 sol the particle dimensions are reduced to a size where practically all the photoinduced charge carriers attain the particle surface before recombination can occur [143]. Contrary to the observations of these authors, Stramet and Thomas reported that they could not promote reactions such as I^- oxidation by irradiating directly into the Fe_2O_3 absorption band [193]. They found that excited tris bipyridine, $Ru(bpy)_3^{2+*}$, is quenched by amorphous Fe_2O_3 and that MV^+ causes the dissolution of this colloid.

The photogeneration of hydrogen from a water-methanol mixture using Nb-doped and Mg-doped Fe_2O_3 powder suspensions has been examined by Somorjai and coworkers [194]. The reaction was driven mainly by bandgap (2.2 eV) radiation. However, small amounts of H_2 were also formed upon sub-bandgap illumination.

WO_3, which has a band gap of 2.6 eV, has not yet been used as colloid in photochemical studies, although a few experiments with suspensions have been reported. When acidic WO_3 suspensions containing Fe^{3+} ions are illuminated with blue light, oxygen is produced [195]. A small deposit of RuO_2 on the WO_3 particles increases the O_2 yield moderately. Fe^{3+} acts as an acceptor for the photo-produced electrons in WO_3, the positive holes thus being able to oxidize water. O_2 was not produced in the absence of Fe^{3+}. The Fe^{2+} ions produced inhibit the O_2 formation at longer illumination times as they act as hole scavengers. Ag^+ ions are a better acceptor of the electrons than Fe^{3+} [196]. The photo reaction leads to silver metal, and O_2 generation is sustained until all the silver is reduced. Another reaction that was observed was the simultaneous oxidation of Cl^- to Cl_2 and reduction of O_2 in WO_3 suspensions containing O_2 and Cl^- ions [197].

Radiation chemical studies were carried out with an acidic $WO_3 \cdot H_2O$ sol stabilized by polyvinyl alcohol [198]. It was found that $(CH_3)_2COH$ radicals inject electrons into the colloidal particles. A long-lived blue color arose and the absorption spectrum showed a rising absorption above 700 nm. This absorption could have been produced by free electrons, although it could not be ruled out that the electrons reduced H^+ ions to produce tungsten bronze. Addition of reducible compounds such as O_2 or Fe^{3+} to the blue sol led to the disappearance of the blue color. In the presence of Cu^{2+} ions, the equilibrium $Cu^+ \rightleftarrows Cu^{2+} + e^-(WO_3)$ was established.

Although they are not colloids in the usual sense, heteropolytungstates may briefly be mentioned here as photocatalysts. In an illuminated solution containing $SiW_{12}O_{40}^{4-}$, methanol, and colloidal platinum, hydrogen is produced. The polyanion is excited by light and an electron is transferred from a methanol molecule to produce a $\dot{C}H_2OH$ radical and a $SiW_{12}O_{40}^{5-}$ anion. The CH_2OH radical in turn reduces another $SiW_{12}O_{40}^{4-}$ ion. The reduced polyanions then reduce water in a Pt-catalysed electron transfer reaction [199, 200].

Photoreactions on ZnO powder in aqueous suspension and in contact with gases have often been studied during the last few decades, and only a few aspects of this work are reviewed here. For example, nitrous oxide and methyl iodide were found to decompose when brought into contact at 20 °C with the illuminated surface of ZnO [201, 202], and nitrate, indigo carmine and p-nitrosodimethylaniline were found to be reduced in aqueous suspensions [203]. ZnO is of special interest as it is one of the standard electrode materials in conventional semiconductor electrochemistry and photoelectrochemistry [204–207]. Colloidal ZnO has not been available until recently. It can be prepared by precipitating Zn^{2+} with NaOH in alcoholic solution, making use of the dehydrating properties of alcohol to prevent the formation of the hydroxide.

The colloid can be recovered in the form of a powder with a specific surface of more than 100 m^2/g and can be redissolved in alcohol or water (pH > 7) [208, 209].

The generation of O_2 in suspensions of ZnO containing Ag^+ ions, which has been known for a long time [125], has again been studied in recent years [210, 211], and experiments on colloidal solutions have also been carried out in our laboratory. Hada et al. measured the quantum yield of silver formation as a function of Ag^+ concentration [210]. They found that the quantum yield φ could be expressed as

$$\varphi^{-1} = \frac{k_2}{\alpha k_1 \cdot \theta} [Ag^+]^{-1/n} + \theta^{-1} \quad (57)$$

where α and n are the constants in Freundlich's adsorption formula, θ the excitation efficiency for formation of conduction band electrons, k_1 the rate constant of reaction of conduction band electrons with adsorbed Ag^+ ions, and k_2 the rate constant for competing reactions of the electrons such as trapping and recombination. The amount of O_2 corresponded to the stoichiometrically expected value. In our experiments with colloidal solutions containing 10^{-4} M ZnO, twice as much O_2 was developed during a sufficiently long irradiation time than was present in ZnO, which shows that the oxygen essentially originates from water and not from ZnO.

ZnO (suspension) sensitizes the photoreduction of Ag^+ by xanthene dyes such as uranin and rhodamine B. In this reaction, ZnO plays the role of a medium to facilitate the efficient electron transfer from excited dye molecules to Ag^+ adsorbed on the surface. The electron is transferred into the conduction band of ZnO and from there it reacts with Ag^+. In homogeneous solution, the transfer of an electron from the excited dye has little driving force as the potential of the Ag^+/Ag^0 system is −1.8 V [47] (Sect. 2.3). It seems that sufficient binding energy of the silver atom formed is available in the reduction of adsorbed Ag^+ ions, i.e. the redox potential of the silver couple is more positive under these circumstances.

Figure 28 (unbroken line) shows the absorption and fluorescence spectra of colloidal ZnO in aqueous solution at pH = 11.7 [208]. The weak fluorescence band at wavelengths close to the onset of absorption was attributed to the fluorescence of the

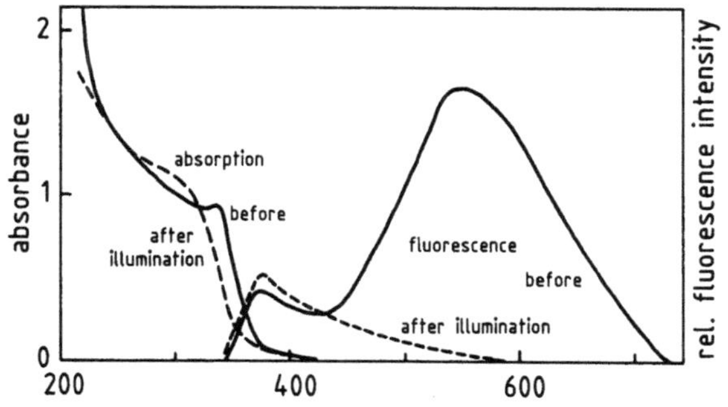

Fig. 28. Absorption spectrum and fluorescence spectrum of a ZnO sol in water, pH = 11.7, before and after 330 nm illumination [208]

exciton, while the broad band was explained by recombination fluorescence of trapped charge carriers. The lifetimes of these two fluorescences were <1 ns and 10 ns, respectively. 10^{-5} M Cu^{2+} quenched the broad fluorescence band at 540 nm but did not influence the 370 nm fluorescence. Figure 28 (dashed) also shows the absorption and fluorescence spectra after a 1 min illumination of the deaerated sol with 330 nm light. The band at 540 nm has disappeared and that at 370 nm is slightly increased. After the solution had been left to stand for 30 min, the 540 nm band was present again. Immediate recovery of this band occurred upon the admission of air to the illuminated solution. It is also seen from the figure that the absorption spectrum of ZnO is shifted to shorter wavelengths after the illumination. This shift is recovered upon standing for 30 min or upon the admission of air.

These effects were explained in terms of the Hauffe mechanism [206] for the dissolution of ZnO:

$$ZnO_s \xrightarrow{h\nu} Zn_s^+ + O_s^- \tag{58}$$

$$Zn_s^+ \rightarrow Zn_{aq}^{2+} + e^- \tag{59}$$

$$e^- + O_s^- + H_2O \rightarrow 2\,OH_{aq}^- \tag{60}$$

$$Zn_{aq}^{2+} + 2\,OH_{aq}^- \rightarrow ZnO_s + H_2O \tag{61}$$

where the subscript "s" indicates surface species. The Zn_s^+ centers quench the fluorescence at 540 nm by reacting with positive holes: $Zn^+ + h^+ \rightarrow Zn^{2+}$, the electrons simultaneously formed by light absorption producing new Zn^+ centers. After illumination, reactions 59–61, which are rather slow, take place and lead to the restoration of the material. When oxygen is admitted after illumination, the restoration of ZnO occurs rapidly via the reactions

$$Zn_s^+ + O_2 \rightarrow Zn_s^{2+} + O_2^- \tag{62}$$

$$O_2^- + O_s^- \rightarrow O^{2-} + O_2 \tag{63}$$

$$Zn_s^+ + O_s^- \rightarrow ZnO \tag{64}$$

Instead of postulating Zn_s^+ as intermediate, as it has a highly negative potential and is possibly unstable in ZnO, one may write the above mechanism with $Zn^{2+}e^-$ pairs. The blue-shift in the absorption upon illumination was explained by the decrease in particle size. The Hauffe mechanism was abandoned after it was recognized that an excess electron on a colloidal particle causes a blue-shift of the absorption threshold (see Fig. 19). In fact, in a more recent study it was shown that the blue-shift is also produced in the electron transfer from CH_2OH radicals to colloidal ZnO particles [144a].

When deaerated propanol-2 solutions of colloidal ZnO were irradiated for longer times, a black precipitate of Zn metal was formed. In the presence of 10^{-4} M methyl viologen in the alcohol solution, MV^+ was produced with a quantum yield of 80% [208].

The technique of spin trapping was used to study the free radicals formed in the illumination of aqueous ZnO suspensions [212]. OH radicals could be detected. In the absence of O_2 in the solution the OH yield was smaller. When O_2 was present, it was consumed and H_2O_2 was formed. In the presence of formic acid, the CO_2^- radical was detected. It was formed either via $OH + HCO_2^- \rightarrow H_2O + CO_2^-$ or via the oxidation of formate by positive holes.

Cadmium oxide, CdO, is a semiconductor with a band gap of 2.3 eV. Irradiation of CdO powder suspended in alkaline solution resulted in the formation of O_2 when an electron acceptor such as ferricyanide was present in the solution. When RuO_2 was deposited onto the surface of the CdO particles the yield of O_2 decreased relative to naked CdO. In this respect CdO differs from TiO_2 where RuO_2 is mandatory if O_2 evolution is to be observed [213]. Colloidal CdO has not been known until recently. It can be made by precipitating $Cd(OH)_2$ in a water-methanol mixture, drying the compound and slowly heating it in vacuum until a pale yellow powder remains. This powder had a broad particle size distribution, the BET surface area being several 10 m^2/g. It can be redissolved in alcohol or water to give weakly yellow solutions. The colloid is stable in the alcoholic solution but hydrolyses slowly in aqueous solution to give the hydroxide. The yellow solutions are photoactive. Silver ions are reduced upon illumination and methyl viologen is reduced at pHs above 11 [214].

In_2O_3 has photochemical properties similar to CdO. The light absorption of an In_2O_3 powder suspended in water commences below 500 nm and a band gap of 2.5 eV was derived from the spectrum. In the presence of Ag^+ ions in the solution, illumination leads to oxygen formation, the efficiency being about 10 times lower than for O_2 production on WO_3 powder [215]. In_2O_3 has not been available as a colloid until recently. It has recently been made in the author's laboratory by the method mentioned above for colloidal CdO. Illumination of colloidal CdO and In_2O_3 leads to a blue-shift of the absorption edge, although the effect is not as pronounced as in the case of ZnO [214].

Manganese dioxide is used as an oxidizing agent and as a catalyst for redox processes such as the decomposition of peroxides. The mechanism of such reactions was studied using transparent colloidal solutions. Colloidal MnO_2 of 4 nm mean particle size was made by exposing a $KMnO_4$ solution to γ-radiation until the conversion of the permanganate was complete. The reactions of $(CH_3)_2COH$ radicals (see footnote, page 5) and O_2^- radicals with the colloidal particles were studied by pulse radiolysis [216] and those of H_2O_2 by stop flow techniques [217]. It was found that the radicals and H_2O_2 transfer electrons to MnO_2. The mechanism of the O_2^- reaction was formulated as follows:

$$O_2^- + Mn_s^{4+} \rightarrow O_2 + Mn_s^{3+} \tag{65}$$

$$O_2^- + Mn_s^{3+} \rightarrow O_2 + Mn_s^{2+} \tag{66}$$

$$Mn_s^{2+} + Mn_s^{4+} \rightarrow 2\,Mn_s^{3+} \tag{67}$$

(s: surface species)

Most interesting is the co-proportionation reaction 67, which is responsible for the accumulation of Mn_s^{3+} centers. The Mn_s^{3+} centers react much faster with O_2^- than the Mn_s^{4+} centers. The reaction is therefore autocatalytic. When about 70% conversion of Mn_s^{4+} to Mn_s^{3+} is reached, reaction 67 is slowed down, and O_2^- can reoxidize Mn_s^{2+}

centers, being itself reduced to H_2O_2. At this stage of the process, the catalysed disporportionation $2\,O_2^- \xrightarrow{2H^+} H_2O_2 + O_2$ takes place. The reaction of H_2O_2 with MnO_2 follows a similar scheme. At the start, the colloid is activated by production of Mn_s^{3+} centers and oxidation of H_2O_2 to $2\,H^+ + O_2$. After activation of the catalyst, the catalysed disproportionation $H_2O_2 \rightarrow H_2O + 1/2\,O_2$ takes place.

5 Size Quantization Effects in Semiconductor Particles

5.1 General Remarks

Differences in the absorption spectra of colloidal and macrocrystalline semiconductors were first recognized for CdS [218] and AgBr [219]. The absorption of 3 nm particles of CdS in aqueous solution begins close to 515 nm, the wavelength at which bulk CdS starts to absorb; however, the increase in absorption at shorter wavelengths is much less steep than for the macrocrystalline material (Fig. 6). The effect was first explained by a possible amorphous structure of the colloidal particles [8]. However, after it was shown by Brus and co-workers [220] that the particles had an ordered structure, it became clear that the differences in absorption were due to the small particle size. Very small CdS particles start to absorb at wavelengths distinctly shorter than 515 nm. This was observed in two simple experiments. Brus [220] found that the threshold of absorption shifted towards longer wavelengths when a freshly prepared colloid of CdS aged. In our experiments on the photodegradation of CdS and ZnS particles, a blue shift of the threshold was observed (Figs. 6 and 7). After these first observations on size quantization effects, methods for the preparation of extremely small particles were developed to investigate these effects under more clearly defined conditions. In these methods, the solution of a cadmium salt is rapidly mixed with H_2S in the absence of air. It is useful to choose an organic solvent such as acetonitrile [220] or alcohol [71, 221] as the reaction medium. CdS has a low solubility in these solvents, and this prevents the particles from rapid Ostwald ripening. Precipitation at -40 °C also favors a small particle size. Precipitation in aqueous solutions is carried out in the presence of sodium polyphosphate which complexes the particles and in this way decreases the solubility of CdS. The concentration of polyphosphate should not exceed the concentration of CdS formed by very much. Sols of complexed particles can be evaporated and the colloid recovered in the form of a powder of high specific surface ($>100\ m^2/g$). The powder can be redissolved to give a transparent sol [71].

The size effects are generally described by the well-known quantum mechanics of a "particle in a box". The electron and the positive hole are confined to potential wells of small dimension and this leads to a quantization of the energy levels (which in the bulk material constitute virtual continua in the conduction and valence band, respectively). The phenomena arise when the size of the colloidal particle becomes comparable to the DeBroglie wavelength of the charge carriers. The quantization effects for an electron in an evacuated box become significant at box dimensions of some 0.1 nm. However, in the colloidal particles the effects can already be seen at a much larger particle size. The reason for this lies in the fact that the effective mass of a charge carrier, which moves in the periodic array of the constituents of the crystal lattice, is generally much lower than the mass of an electron in free space. This results in a larger DeBroglie wavelength. The smaller the effective mass of the charge carriers, the more pronounced are the optical size effects. The effects can lead to drastic changes in the color of a material, the color of its luminescence, and its catalytic properties. To indicate that a material has unusual properties as compared to those of macrocrystalline material we proposed using the prefix Q before the chemical formula [71].

Semiconducting systems of low dimensionality, so-called superlattices, are frequently used in the field of micro-electronics [222–226]. In one configuration, these structures consist of alternating thin layers of two semiconductors with two different band gaps, the small gap semiconductor forming a series of square-well potentials and the large gap semiconductor forming a series of potential barriers. Potential wells for both conduction band electrons and valence band holes are formed, and the position of these bands depends on the thickness of the wells and barriers. As the quantization occurs only in one dimension in these structures, the shifts of the electronic levels are only of the order of 0.1 eV, while the shifts in the colloidal particles, where quantization is operative in all three dimensions, can amount to several eV. A superlattice electrode consisting of GaAs as the low band gap material and $GaAs_{0.5}P_{0.5}$ as the large band gap material was recently used in photoelectrochemical experiments [227]. Pronounced structures were observed in the photocurrent spectrum. Two-dimentional systems called quantum wires have also been prepared in solid state physics [228, 229]. Solid state physicists often number the dimensions in which there is no quantization. Colloidal particles represent the zero-dimensional case in this nomenclature. It may finally be mentioned that size quantization effects in the optical spectrum of CdS were also observed for tiny crystals grown in a silicate glass matrix at high temperatire [230] and for CdS and PbS clusters encapsulated in zeolites [230 a)]. Q-CdS is also formed when surfactant vesicles are used as hosts for the semiconductor particles [231, 233].

It was also reported [234] that microcrystallites of layered semiconductor PbI_2 were prepared in colloidal form. The spectrum of such a colloidal solution consisted of three absorption bands in the UV which were considerably bueshifted from the absorption threshold of macrocrystalline PbI_2. These results were explained by carrier confinement in three differently sized crystallites, each a single layer (~7 Å) thick. However, complexes of PbI_2 with iodide have similar absorption bands, and it seems at the present time that additional experiments have to be carried out to ascertain the colloidal nature of the absorbing species. Size quantization was also reported for colloids of red HgI_2 in acetonitrile [234 a)].

5.2 Absorption of Q-CdS and Some Other Materials

The absorption spectrum of CdS of very small particle size contains a maximum at wavelengths shortly below the beginning of absorption and, depending on the method of preparation, additional maxima at shorter wavelengths [71]. Figure 29 shows a typical example. As the sol ages intensity variations occur in these maxima, the onset of absorption is shifted towards longer wavelengths and new mayima appear in the long wavelength range, while the maxima at short wavelengths disappear. The position of a maximum is not changed during aging.

Slightly below the lower edge of the conduction band of a semiconductor are the levels of the exciton, i.e. a state where the charge carriers do not move independently

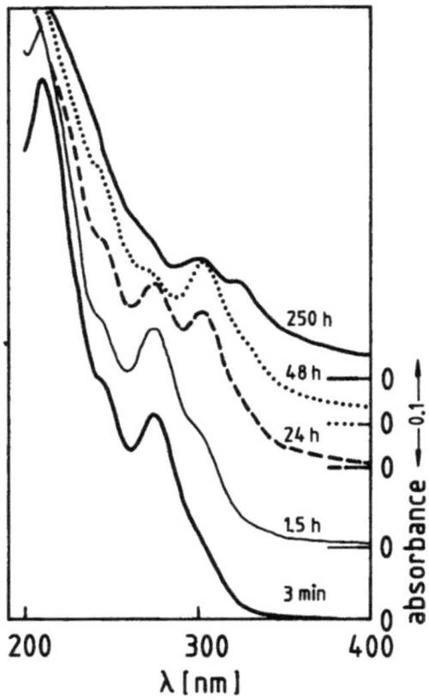

Fig. 29. Absorption spectrum of Q-CdS at various time periods after preparation. 2×10^{-4} M CdS and 2×10^{-4} M $Na_6(PO_3)_6$. Spectra shifted in vertical direction, zero line indicated on the right ordinate axis

of each other. The electron is in an orbital around the hole, like that around the proton in a hydrogen atom. The Coulombic attraction between the two charges is strongly decreased by the dielectric properties of the medium. An exciton in macrocrystalline CdS has a diameter of about 5 nm and its binding energy is only 0.03 eV. Between the lowest state of the exciton and the lower edge of the conduction band exists a Rydberg series of energy levels of the exciton. As the lowest exciton state is so close to the conduction band, thermal activation often leads to the separation of the charge carriers and the optical transitions to the exciton state cannot be resolved as separated bands in the absorption spectrum if it is recorded at ambient temperature. In the small particles the levels of the exciton and of the conduction band are shifted to higher energies [71,235,236], the difference between the exciton level and the lower edge of the conduction band becoming larger with decreasing particle size. This is the reason why the optical transitions to the exciton state produce an absorption band which can be resolved at least partially in the spectrophotometric recordings at ambient temperature. The smaller the particles, the shorter is the wavelength of this absorption maximum. As the particles have a diameter smaller than that of an exciton in macrocrystalline material, the excitons are highly distorted. It is known from solid state physics that excitons can have different energies and shapes depending on the material conditions under which they are formed [237].

The question now arises how the various maxima that are sometimes observed can be explained. Two explanations seem to be possible. One could ascribe the maxima to transitions to different electronic states of the small particles [230]. In this case, the structured absorption spectrum would only be seen if the sample had a narrow size distribution and one would expect a maximum to change position upon particle growth by aging.

In the second kind of explanation, it was postulated that particles of certain sizes are particularly abundant in the size distribution [71]. The occurrence of "magic" agglomeration numbers was explained by the following mechanism of precipitation:

$$m_0Cd^{2+} + m_0S^{2-} \rightarrow (CdS)_{m0} \quad (68)$$

$$(CdS)_{m0} \rightarrow (CdS)_{m0-1} + Cd^{2+} + S^{2-} \quad (69)$$

$$Cd^{2+} + S^{2-} + (CdS)_{m+n} \rightarrow (CdS)_{m+n+1} \quad (70)$$

$$(CdS)_{m0} + (CdS)_{m0} \rightarrow (CdS)_{2m0} \quad (71)$$

Rapid mixing of Cd^{2+} and S^{2-} solutions leads to a very small colloid, the size distribution of which peaks at the agglomeration number m_0 Eq. (68). This primary size distribution rapidly rearranges as the particles become larger either by Ostwald ripening Eqs. (69) and (70) or by association Eq. (71). Ostwald ripening would not lead to a structured size distribution. However, growth by association would produce "magic" numbers, as the most abundant number, m_0, of the primary size distribution would be preserved in the form of integer multiples. Structured size distributions have not yet been seen in colloid chemistry. They cannot be expected to exist for the usual colloids of large size, as these colloids are generally grown gradually from small seeds. The fact that, in the range of Q-CdS, particles of different size can be distinguished by their optical absorption makes it possible to recognize structured size distributions. Unfortunately, it has not yet been possible to prove the hypothesis of "magic" numbers by determining the size distribution in the electron microscope. CdS particles down to 1 nm can be seen in the electron microscope, but the contrast in the pictures is weak and does not permit the resolution of the fine structure of the size distribution.

Figure 30 shows the results of an experiment in which a solution of Q-CdS was fractionated by exclusion chromatography in a column of sephacryl-gel [238]. This column material has holes which the smaller particles penetrate and reside in for some time. The first fraction therefore contains the larger particles. The upper part of the figure shows the absorption spectrum of the starting material, and the lower part the spectra of six fractions. The first fraction has an unstructured spectrum beginning at

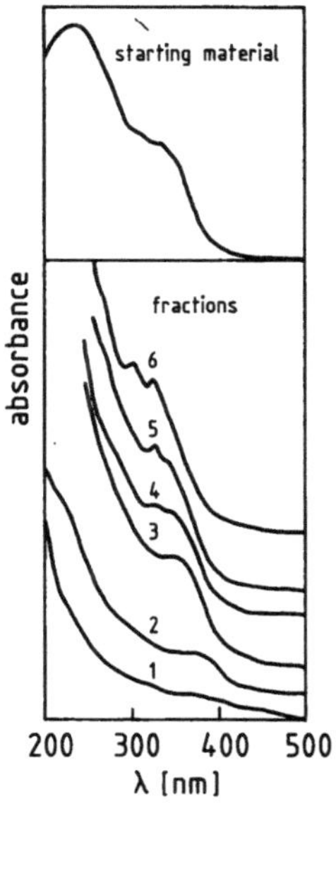

Fig. 30. Absorption spectrum of a Q-CdS sol (upper part) and of six fractions (lower part). Spectra are vertically shifted [238]

500 nm. The second fraction already has a shoulder at 390 nm which is attributed to an exciton transition. The shoulder grows into a maximum and is shifted towards shorter wavelengths in the next fractions. Also note that the maxima present in a fraction disappear or become less pronounced in the succeeding fractions, while new maxima at shorter wavelengths become more important. These findings clearly show that an enrichment of particles of different size occurred in the various fractions.

The structured absorption spectra are often observed when sodium hexametaphosphate is used as a stabilizer. The commercial samples do not contain $Na_6(PO_3)_6$, but polyphosphates up to a chain length of 450. The development with time of the absorption spectrum of Q-CdS in such solutions was followed using the stop flow method [238]. It was found that structured spectra developed best in solutions that contained $(NaPO_3)_n$ and CdS in the ratio 0.5 to 1 and also contained excess Cd^{2+} ions. It was suggested that precipitation occurred in the bulk solution to give the primary size distribution with the peak at m_0 Eq. (68) and that the small CdS particles then agglomerated Eq. (71) on the polyphosphate matrix. CdS samples with structured absorption spectra were also obtained from precipitation in alcohol solution at −40 °C. Changes took place on warming until a spectrum with only one pronounced maximum remained [71].

Similar changes in the absorption spectrum were observed for colloidal ZnS. A colloid in methanol solution at −77 °C had a maximum at 228 nm. Upon aging, this maximum became weaker and a new one arose at 243 nm. Further aging at room temperature led to a spectrum with one maximum at 265 nm. It was concluded that the −77 °C synthesis created two very small types of crystallites (<2 nm) identified by the 228 and 243 nm peaks. Upon warming, those crystallites were lost and new crystallites of about 2 nm were formed which persisted at room temperature [239]. Absorption spectra with several maxima were also observed for ZnSe, CdSe and In_2Se_3 stabilized by polyphosphate [240].

Cd_3P_2 and Cd_3As_2 are low band gap semiconductors (0.5 and 0.1 eV, respectively). The bulk materials are black and start to absorb in the infrared. These materials have been prepared as colloids in alkaline solution by precipitation of Cd^{2+} with phosphine [62] and arsine [63]. Depending on the conditions of preparation, particles of different sizes (between about 2 and 10 nm) were obtained, which could also be recovered in the solid state after evaporation of the solvent. The color of these materials ranged from black to colorless with decreasing particle size, with all kinds of intermediate colors in the visible.

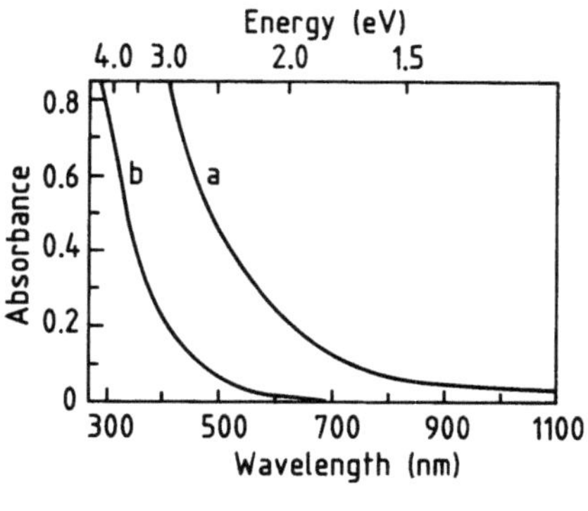

Fig. 31. Optical absorption spectra of PbS colloids (1.0×10^{-4} M) in acetonitrile: a) particle diameter ~50–200 Å; b) particle diameter ~20–30 Å [241]

Strong size quantization effects were found for lead sulfide [241-243]. While the sulfide and selenides of zinc and cadmium are large band gap (>2 eV), sp^3 hybridized materials with diamond lattices and predominantly covalent bonding, PbS is an ionic crystal with a rock salt lattice and an infrared (0.29 eV) band gap. In materials of the latter kind, a shift takes place in the absorption threshold with decreasing particle size, as in the zinc and cadmium salts. However, the absorption spectrum is unstructured and does not contain the size-dependent exciton peak close to the absorption threshold. This can be seen from Fig. 31 in which the spectra of two PbS samples of different mean particle size are shown. Although the spectra lack the near infrared absorbance characteristic of bulk PbS, they have long tails towards longer wavelengths. Besides the usual "particle in a box" effect, the spectrum may be determined by contributions from non-direct optical transitions. It may be that the highest hole state in the small particles is not in the same region of the Brillouin zone as the lowest localized electron state. This would give the spectrum a forbidden origin, as in indirect gap transitions [242]. Similarly unstructured spectra have also been observed for colloids of AgS and Sb_2S_3 in the author's laboratory.

In Q-particles, the electron produced by light absorption should be on a more negative potential and the hole generated should be on a more positive potential than in the macrocrystalline materials. Chemical effects have been observed which can be explained by the stronger reducing properties of the electrons and the stronger oxidizing properties of the holes. For example, H_2 evolution was observed on illuminated PbSe and HgSe colloids and CO_2 reduction on CdSe when the particle size was less than 5 nm. These reactions did not occur on suspended larger particles [243]. Q-CdS particles catalyse the formation of the hydroperoxide $(CH_3)_2C(OH)OOH$ when they are illuminated in aerated propanol-2 solution, while larger particles are not efficient in this respect [71].

5.3 Fluorescence of Q-CdS and Some Other Materials

Particles of different size have not only different absorption spectra but also different fluorescence spectra. In Fig. 32, the absorption spectrum of a Q-CdS sample and two

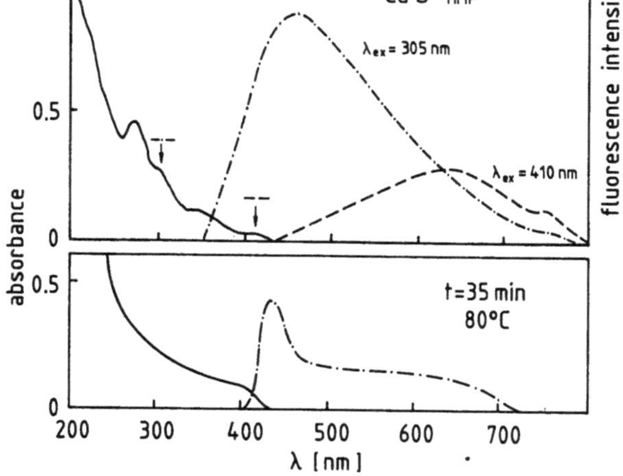

Fig. 32. Absorption spectrum and fluorescence spectra of Q-CdS at two wavelengths of excitation (arrows, upper part) and spectra after aging (lower part)[4,71]

fluorescence spectra are shown for excitation at 305 nm and 410 nm. One observes a broad fluorescence band, the maximum of which shifts towards longer wavelengths with increasing wavelength of the exciting light (i.e. with increasing size of the absorbing particle). When the solution was aged by warming, the absorption spectrum lost its structure. This can also be seen from Fig. 32. In the fluorescence spectrum a narrow band appeared at the onset of absorption besides the broad band at longer wavelengths. The lifetime of the decay of fluorescence in the narrow band was shorter than 1 ns, while it amounted to several 100 ns in the broad band, the decay occurring here in a multi-exponential manner. Small amounts of methylviologen quenched the broad fluorescence band, but not the narrow one. The narrow band was attributed to the direct fluorescence of the exciton-like state which is produced by light absorption, and the broad band to the recombination of charge carriers trapped in anion vacancies at the surface of the particles. In very small particles, the transition from the exciton state to the defect states is so fast that the direct fluorescence does not appear.

The fluorescence excitation spectrum of a Q-CdS sample, with several maxima in the absorption spectrum, also has a number of peaks. However, the maxima in the two spectra do not always occur at the same wavelengths [71]. This effect is not surprising, as excitation at different wavelengths leads to the excitation of particles of different sizes which do not have the same fluorescence intensity at the wavelength where the fluorescence is recorded.

The Q-materials of Cd_3P_2 and Cd_3As_2 mentioned in the preceding section fluoresce strongly both in solution and in the solid state. The fluorescence band, which often shows a structure, has its maximum at a wavelength substantially longer than that of the onset of absorption. 10 nm particles fluoresce mainly in the infrared, 2nm particles fluoresce green. All kinds of fluorescence color were observed for particles within this size range [62,63,240].

5.4 Pulse Radiolysis and Flash Photolysis of Q-CdS

The pulse radiolysis studies described in Sect. 3.7 showed that OH radicals react with colloidal CdS to produce a long-lived product with a broad absorption band in the visible. The band is blue shifted with decreasing particle size as shown by Fig. 20. This phenomenon was explained as a consequence of size quantization. The absorption of the product, which was identified as a surface-trapped positive hole, or, chemically speakting, as an S^- radical anion, is brought about by the transfer of an electron into higher unoccupied levels of the colloidal particle. As these levels shift to higher energies with decreasing particle size, the absorption of the hole occurs at shorter wavelengths. A similar shift with particle size was observed for the absorption band of the electron deficient surface state that was produced in the reaction of OH with Cd_3P_2 [124].

In addition to this size quantization effect, a second effect was observed in the reaction of OH with Q-CdS. As is seen from Fig. 20, oscillations occur in the difference spectrum of the product at wavelengths below the absorption threshold. The absorptions in this wavelength range are not caused by the positive holes but are due to the changes in size of the colloidal particles upon OH attack. The absorption spectrum shifts slightly towards shorter wavelengths as the particles become smaller in reacting with OH. The maxima in the spectrum before and after OH attack are not positioned at the same wavelengths but are slightly displaced, and this leads to the oscillatory behavior of the difference spectrum [124].

Laser flash experiments were also carried out with Q-CdS sols, in which the emission of hydrated electrons was observed [110]. The quantum yield was significantly greater than in similar experiments with larger particles of yellow CdS (Sect. 3.7). The electron emission was attributed to the interaction of two excitonic states in a particle produced during the flash: $CdS(e^- - h^+)_2 \rightarrow CdS(h^+) + e_{aq}^-$. The emitted electrons disappeared after the laser flash within 10 μs. After this time a long-lived absorption remained which was identical with the above-mentioned absorption of holes produced by OH radicals in the pulse radiolysis experiment.

5.5 Metal Oxides

Size quantization effects in colloidal ZnO can be observed when the particles grow after their formation (see Sect. 4.6). At first, the colloid fluoresces blue-green, after a few minutes green, and finally yellow-green. The shifts of the fluorescence band can be seen from Fig. 33. The figure also shows that the sol starts to absorb at 310 nm at first, a wavelength substantially lower than the wavelength of the absorption threshold of macrocrystalline ZnO (360 nm; photon energy 3.4 eV). Upon aging, the absorption spectrum is shifted to longer wavelengths until the colloid absorbs like a macrocrystal. The spectra have shoulders about 30 nm below the th,reshold which were attributed to optical transtions to the exciton state [208,209].

The properties of a tungsten oxide monolayer on 0.18 μm silica particles were studied by Leland and Bard [244]. The material was prepared by a controlled WCl_6 hydrolysis technique. These particles differ from the colloidal Q-particles that they are small in only one dimension; thus these layers are related to the semiconductor

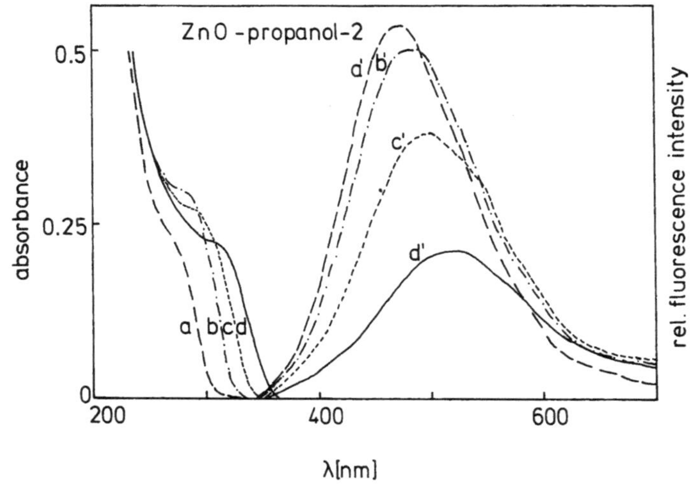

Fig. 33. Absorption und fluorescence spectra of aerated 3×10^{-4} M ZnO sol in propanol-2 at various times of aging [208]. pH = 12. a) 1 min; b) 5 min; c) 20 min; d) 60 min

superlattices (section 5.1). It was found that the material displayed a bandgap which was 0.3 eV higher in energy than that for bulk WO_3.

5.6 Theory

The first calculations of the shift of the band gap in small particles of a semiconductor were made by Efros und Efros [233]. They treated the model of two Coloumb-interacting particles confined to a sphere for the limiting cases $R \ll R_{exc}$ and $R \gg R_{exc}$ (where $R_{exc} = \hbar^2\varepsilon/\mu e^2$ is the radius of an exciton in the macrocrystal) and suggested a Born-Oppenheimer-like approximation for other values of R. A full wave-mechanical treatment of the problem was reported by Brus [235,236], who calculated the lowest eigenstate of an exciton by solving Schrödinger's equation at the same level of approximation as is generally used in the analysis of bulk crystalline electron-hole states. The exciton wave function was approximated by one or a few configurations $\Psi_i(r_e) \cdot \Psi_j(r_h)$ of particle-in-a-spherical-box orbitals, and using the usual values of the effective masses m_e and m_h of the charge carriers. Fig. 34 gives the wavelength of the onset of absorption as a function of the diameter, 2R, of CdS particles. Curve c represents Brus's results.

In other models, the confined exciton was treated as one electron with the reduced mass $\mu = (1/m_e + 1/m_h)^{-1}$ which moves in the field of the walls, and a positive hole h^+ fixed at the centre of the sphere. Curve a in Fig. 34 gives the lowest energy level (wavelength of transition to this level) as calculated in the semi-classical approxima-

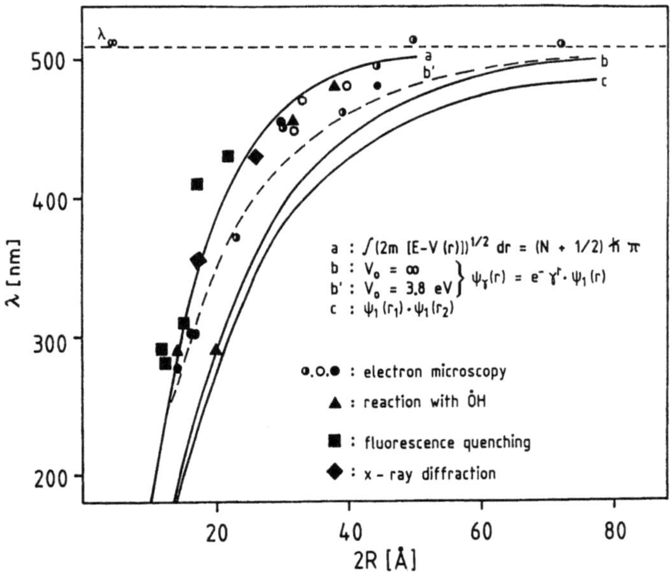

Fig. 34. Wavelength of threshold as a function of diameter of CdS particles. λ_∞ : macrocrystal; points: experimental data; curves: calculated [245]

tion [71]. In this approximation the eigenvalues were estimated which satisfy the condition

$$J = \int_{r_1}^{r_2} \frac{[2\mu^*(E - V^{(r)})]^{1/2}}{\hbar} dr = \left(n + \frac{1}{2}\right)\pi \tag{72}$$

(J = phase integral) with the potential function

$$V_{(r)} = -\frac{e^2}{\epsilon r} + \frac{\hbar^2(1 + 1/2)^2}{2\mu^* r^2} \tag{73}$$

(μ^* = reduced effective mass, ϵ = dielectric constant of CdS, 1 = quantum number of angular momentum. r_1 and r_2 are the roots of $V_{(r)}$). For the same model, an accurate quantum mechanical solution was obtained using a wave function of the form $\exp(-\gamma r)\,\Psi_1(r)$ where the hydrogen-like factor takes account of the Coulombic attraction and $\Psi_1(r)$ is the lowest particle-in-a-spherical-bos orbital [245]. The result is shown by curve b in Fig. 34.

All of these one- and two-body models have assumed hard walls for the box (potential $V = \infty$ for $r > R$). The actual potential energy difference between the lower edge of the conduction band of the macrocrystal and the vacuum level amounts to 3.8 eV. This potential depth was used in the quantum mechanical calculation of curve b′. It is seen that the energy lowering is substantial, particularly at small diameters.

These curves may now be compared to the experimental results which are described by the various points in Fig. 34. CdS samples of different particle size were prepared and the absorption spectra measured. The size of the particles was determined by the four methods mentioned in the figure. The difficulties that arose in these measurements were due to the fact that it was not possible to prepare monodisperse samples of CdS. Fig. 35 shows a typical size distribution and the absorption spectrum of this sample. Some points in Fig. 34 were obtained by correlating the maximum A in the size distribution with the wavelength at which the exciton maximum A′ in the spectrum appeared. Other points were obtained by correlating the extrapolated values B and B′ in Fig. 35. The reaction of OH radicals, which was described in Sect. 3.7 (see Eq. 48), was also used to obtain the mean size of the particles. Fluorescence quenching with methylviologen (Sect. 3.3) was also applied. As one adsorbed MV^{2+} per colloidal particle is sufficient for quenching, the concentration of colloidal particles can be determined, and from this concentration the size can be derived. In the case of solid CdS particles, X-ray diffraction was used. The diffractogram showed the pattern of cubic CdS. The reflexes [111] and [002] at $2\theta = 27.0$ and $31.03°$ were observed at slightly larger angles than in the macrocrystalline material. The line widening of the diffraction signals was used to calculate the particle diameter [245].

These different methods produced consistent results, as can be seen from Fig. 34. Obviously, in the hard-box model for the confined exciton (curves c and b) the increase in the band gap with decreasing crystallite size is exaggerated. The more realistic assumption of a *finite* potential energy step at the crystallite surface (curve b′) improves the model considerably. The good agreement between the semi-classical curve (a)

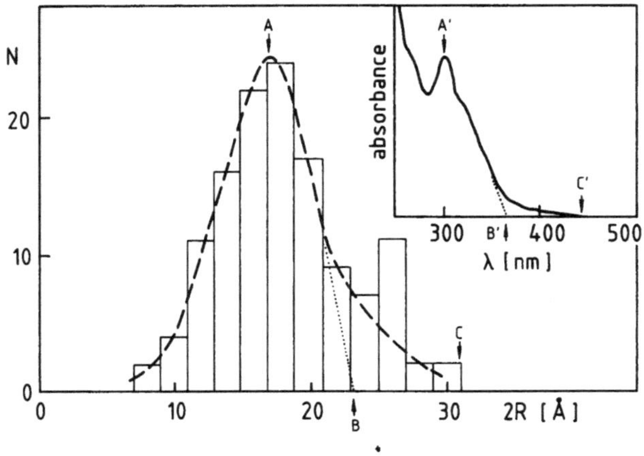

Fig. 35. Size distribution as determined by electron microscopy and absorption spectrum of a CdS sample of small particle size [245]

and the experiments must be partly fortuitous. While the semi-classical approximation gives the exact energy levels for the free exciton, it underestimates the kinetic energy of an electron between infinitely high wells. It could be that this error is compensated by the error introduced by the infinite hard-well assumption. Finally, it should be mentioned that the effective mass is not constant but may increase with decreasing particle size. No exact method seems to exist for taking such an effect into consideration.

Two more accurate quantum mechanical calculations of the two body system have appeared using configuration interaction and Hyllareas functions, as well as a perturbation expansion in R, but still a hard wall potential [245,246]. The term scheme in

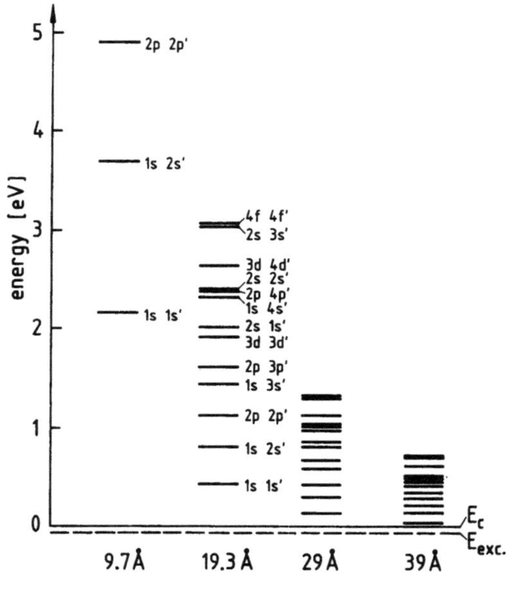

Fig. 36. Energy levels of excitonic states in CdS particles of various radii. Zero: position of the lower edge E_c of the conduction band in macrocrystalline CdS. Exc: Energy of an exciton in macrocrystalline CdS. Effective masses of electrons and holes 0.19 m_0 and 0.8 m_0 respectively. The letters with a prime designate the quantum state of the hole [246]

Fig. 36 resulted from these calculations. Only levels with a total angular momentum number L = 0 are contained in the terms chemes. It is seen that all levels are shifted to higher energies with decreasing particle size and the density of states is decreased. The shift of the lowest level would be slightly less for a finite potential as discussed above, while the higher levels would be influenced more drastically. However, despite these approximations the figure gives an instructive illustration of what happens to the electronic structure of a semiconductor in the transition range between semiconductor and molecular properties.

6 References

1. Bennemann, K. H., Koutecký, J. (eds.): Small particles and inorganic clusters, Proc. 3rd int. meeting, Surface Science Vol. 15 (part 1 and 2) North-Holland Amsterdam 1985
2. Ostwald, Wo.: Die Welt der vernachlässigten Dimensionen, 4. Aufl., Th. Steinkopff, Dresden 1920
3. Henglein, A.: in Photochemical conversion and storage of solar energy (ed. Rabani, J.) The Weizmann Science Press of Israel, part A, 115, 1982
4. Henglein, A.: in Modern trends of colloid science in chemistry and biology (ed. Eicke, H.-F.) Birkhäuser, Basel, 126, 1985
5. Baxendale, J. H., Busi, F. (eds.): The study of fast processes and transient species by electron pulse radiolysis, D. Reidel Publishing Co., Dordrecht 1982
6. Henglein, A., Schnabel, W., Wendenburg, J.: Einführung in die Strahlenchemie, Verlag Chemie, Weinheim 1969
7. Wilets, F. W.: in Progress in Reaction Kinetics, Vol. 6, part 1 (ed. Jennings, K. R., Cundall, R. B.) Pergamon Press, Oxford 1971
8. Henglein, A.: Ber. Bunsenges. Phys. Chem. *86*, 301 (1982)
9. Henglein, A.: in Electroanalytical Chemistry (ed. Bard, A. J.) Marcel Dekker, Inc., New York and Basel, 9, 163, 1976;
   Toffel, P., Henglein, A.: Faraday Disc. *63*, 124 (1978)
10. Henglein, A.: J. Phys. Chem. *83*, 2209 (1979)
11. Grätzel, C. K., Grätzel, M.: J. Am. Chem. Soc. *101*, 7741 (1979)
12. Meisel, D.: ibid. *101*, 6133 (1979)
13. Henglein, A.: J. Phys. Chem. *83*, 2858 (1979)
14. Henglein, A.: Ber. Bunsenges. Phys. Chem. *86*, 241 (1982)
15. Bredig, G.: Anorganische Fermente, Leipzig 1901
16. Hückel, W.: Katalyse mit kolloidalen Metallen, Leipzig 1927
17. Spiro, M.: J. Chem. Soc. 3678, 1960; followed by a series of papers on heterogeneous catalysis
18. Spiro, M.: J. Chem. Soc. Faraday Trans. I, 75 (1979)
19. Henglein, A.: J. Phys. Chem. *84*, 3461 (1980)
20. Henglein, A., Lilie, J.: J. Am. Chem. Soc. *103*, 1059 (1981)
21. Henglein, A., Lindig, B., Westerhausen, J.: J. Phys. Chem. *85*, 1627 (1981)
22. Henglein, A.: Ber. Bunsenges. Phys. Chem. *84*, 253 (1980)
23. Lee, P. C., Meisel, D.: J. Catalysis *70*, 160 (1981)
24. Henglein, A., Lilie, J.: J. Phys. Chem. *85*, 1246 (1981)
25. Buxton, G. V., Rhodes, T., Sellers, R. M.: J. Chem. Soc. Faraday Trans. I, *78* 3341 (1982)
26. Mills, G., Henglein, A.: Radiat. Phys. Chem. *26*, 385, 391 (1985)
27. Kopple, K., Meyerstein, D., Meisel, D.: J. Phys. Chem. *84*, 870 (1980);
    Meisel, D., Mulac, W. A., Matheson, M. S.: ibid. *85*, 179 (1981)
28. Westerhausen, J., Henglein, A., Lilie, J.: Ber. Bunsenges. Phys. Chem. *85*, 182 (1981)
29. Duonghong, D., Borgarello, E., Grätzel, M.: J. Am. Chem. Soc. *103*, 4685 (1981)
30. Wagner, C., Traud, W.: Z. Elektrochemie *44*, 391 (1938)
31. Miller, D. S., Bard, A. J., McLendon, G., Ferguson, J.: J. Am. Chem. Soc. *103*, 5336 (1981)
32. Miller, D. S., McLendon, G.: ibid. *103*, 6791 (1981)

33. Delcourt, M.-O., Keghouche, N.: Nouv. J. Chim. *9*, 235 (1985)
34. Albery, W. J., Bartlett, P. N.: J. Electroanal. Chem. *131*, 137 (1982); ibid. *131*, 145 (1982); ibid. *139*, 57 (1982)
35. Albery, W. J., McMahon, A. J.: ibid. *182*, 1 (1985)
36. Albery, W. J., Bartlett, P. N., McMahon, A. J.: ibid. *182*, 7 (1985)
37. Lee, P. C., Matheson, M. S., Meisel, D.: Israel. J. Chem. *22*, 133 (1982)
Matheson, M. S., Lee, P. C., Meisel, D., Pelizzetti, E.: J. Phys. Chem. *87*, 394 (1983)
38. Amouyal, E., Koffi, P.: J. Photochem. *29*, 227 (1985)
39. Venturi, M., Mulazzani, Q. G., Hoffman, M. Z.: J. Phys. Chem. *88*, 912 (1984); Radiat. Phys. Chem. *23*, 229 (1984)
40. Brandeis, M., Nahor, G., Rabani, J.: J. Phys. Chem. *88*, 1615 (1984)
41. Delcourt, M. O., Keghouche, N., Belloni, J.: Nouv. J. Chim. *7*, 1 (1983)
42. Degani, Y., Willner, I.: J. Chem. Soc. Perkin Trans. II, 37 (1986)
43. Delcourt, M. O., Belloni, J., Marignier, J.-L., Mory, C., Colliex, C.: Radiat. Phys. Chem. *23*, 485 (1984)
44. Marignier, J.-L., Belloni, J., Delcourt, M.-O., Chevalier, J. P.: Nature *317*, 344 (1985)
45. Henglein, A., Lindig, B., Westerhausen, J.: Radiat. Phys. Chem. *23*, 199 (1984)
46. Plieth, W. S.: J. Phys. Chem. *86*, 3166 (1982)
47. Henglein, A.: Ber. Bunsenges. Phys. Chem. *81*, 556 (1977)
48. Tausch-Treml, R., Henglein, A., Lilie, J.: Ber. Bunsenges. Phys. Chem. *82*, 1335 (1978)
49. Hilpert, K., Gingerich, K. A.: ibid. *84*, 739 (1980)
50. Henglein, A., Tausch-Treml, R.: J. Colloid. Interface Sci. *80*, 84 (1981)
51. Platz, H., Schenk, P. W.: Angew. Chem. *49*, 822 (1936)
52. Gloor, K.: Helv. Chim. Acta *20*, 853 (1937)
53. Gerischer, H., Willig, F.: in Topics in current chemistry, *61*, 33 (1976)
54. Gerischer, H.: in Advances in electrochemistry and electrochemical engineering (ed. Delahay, P., Tobias, C. W.,) 1, p. 139, New York, Interscience 1961
55. Gerischer, H.: in Solar power and fuels (ed. Bolton, J. R.) Academic Press, Inc., New York 1977, chapter 4
56. Meissner, D., Memming, R., Kastening, B., Bahnemann, D.: Chem. Phys. Lett. *127*, 419 (1986)
57. Meissner, D., Memming, R., Shuben, L., Yesodharan, S., Grätzel, M.: Ber. Bunsenges. Phys. Chem. *89*, 121 (1985)
58. Harbour, J. R., Hair, M. L.: J. Phys. Chem. *81*, 1791 (1977)
59. Harbour, J. R., Wolkow, R., Hair, M. L.: ibid. *85*, 4026 (1981)
60. Henglein, A.: ibid. *86*, 2291 (1982)
61. Henglein, A., Gutiérrez, M.: Ber. Bunsenges. Phys. Chem. *87*, 852 (1983)
62. Weller, H., Fojtik, A., Henglein, A.: Chem. Phys. Lett. *117*, 485 (1985)
63. Fojtik, A., Weller, H., Henglein, A.: ibid. *120*, 552 (1985)
64. Koch, U.: Dissertation, Technische Universität Berlin 1985
65. Curie, D., Prener, J. S.: in Physics and Chemistry of II–VI Compounds (Aven, M., Prener, J. S., eds.) North Holland Publishing Company, Amsterdam 1967
66. Grätzel, M.: in Modern aspects of electrochemistry, (White, R. E., Bockris, J. O. M., Conway, B. E., eds.) Pergamon Press, New York and London 1983
67. Ramsden, J. J., Grätzel, M.: J. Chem. Soc. Faraday Trans. I, *80*, 919 (1984)
68. Kuczynski, J. P., Milosavijevic, B. H., Thomas, J. K.: Phys. Chem. *87*, 3368 (1983)
69. Kuczynski, J., Thomas, J. K.: ibid. *87*, 5498 (1983), Chem. Phys. Lett. *88*, 445 (1982)
70. Weller, H., Koch, U., Gutiérrez, M., Henglein, A.: Ber. Bunsenges. Phys. Chem. *88*, 649 (1984)
71. Fojtik, A., Weller, H., Koch, U., Henglein, A.: ibid. *88*, 969 (1984)
72. Dannhauser, T., O'Neil, M., Johansson, K., Whitten, D., McLendon, G.: J. Phys. Chem. *90*, 6074 (1986)
73. Spanhel, L., Weller, H., Fojtik, A., Henglein, A.: Ber. Bunsenges. Phys. Chem., *91*, 88 (1987)
73a. Spanhel, L., Haase, M., Weller, H., Henglein, A.: J. Amer. Chem. Soc. *109*, 5649 (1987)
73b. Ramsden, J. J., Webber, S. E., Grätzel, M.: J. Phys. Chem. *89*, 2740 (1985)
74. Chestnoy, N., Harris, T. D., Hull, R., Brus, L. E.: J. Phys. Chem. *90*, 3393 (1986)
75. Gutiérrez, M., Henglein, A.: Ber. Bunsenges. Phys. Chem. *87*, 474 (1983)
76. Darwent, J. R., Porter, G.: J. Chem. Soc. Chem. Comm. (1981) 145
77. Darwent, J. R.: J. Chem. Soc. Faraday Trans. II, *77*, 1703 (1981)

78. Kalyanasundaram, K., Borgarello, E., Duonghong, D., Grätzel, M.: Angew. Chem. Int. Ed. Engl. *20*, 987 (1981)
79. Kalyanasundaram, K., Borgarello, E., Grätzel, M.: Helv. Chim. Acta *64*, 362 (1981)
80. Borgarello, E., Kalyanasundaram, K., Grätzel, M.: ibid. *65*, 243 (1982)
81. Borgarello, E., Erbs, W., Grätzel, M.: Nouv. J. Chim. *7*, 195 (1983)
82. Pelizetti, E., Borgarello, E., Serpone, N., Grätzel, M.: in Catalysis on the energy scene (Kaliaguine, S., Mahay, A., eds.) Elsevier Science Publishers, Amsterdam 1984, 327
83. Aruga, T., Domen, K., Naito, S., Onishi, T., Tamaru, K.: Chem. Lett. (Chem. Soc. Japan) (1983), 1037
84. Matsumura, M., Saho, Y., Tsubomura, H.: J. Phys. Chem. *87*, 3807 (1983)
84a. Matsumura, M., Hiramoto, M., Iehara, T., Tsubomura, H.: J. Phys. Chem. *88*, 248 (1984)
85. Rajh, T., Mićić, O. I.: Bull. Soc. Chim. Beograd *48*, 335 (1983)
86. Bühler, N., Meier, K., Reber, J.-F.: J. Phys. Chem. *88*, 3261 (1984)
87. Thewissen, D. H. M. W., Tinnemans, A. H. A., Eeuwhorst-Reinten, M., Timmer, K., Mackor, A.: Nouv. J. Chim. *7*, 191 (1983)
88. Thewissen, D. H. M. W., van der Zouwen-Assink, E. A., Timmer, K., Tinnemans, A. H. A., Mackor, A.: J. Chem. Soc. Chem. Commun. (1984) 941
88a. Naman, S. A., Aliwi, S. M., Al-Emara, K.: Int. J. Hydrogen Energy, *11*, 33 (1986)
89. Matsumura, M., Furukawa, S., Saho, Y., Tsubumura, H.: J. Phys. Chem. *89*, 1327 (1985)
90. Tricot, Y. M., Fendler, J. H.: J. Amer. Chem. Soc. *106*, 2475 (1984); *106*, 7359 (1984)
Rafaeloff, R., Tricot, Y.-M., Nome, F., Fendler, J. H.: J. Phys. Chem. *89*, 533 (1985)
Tricot, Y.-M., Emeren, A., Fendler, J. H.: J. Phys. Chem. *89*, 4721 (1985)
91. Fendler, J. H.: ibid. *89*, 2730 (1985)
92. Reber, J.-F., Rusek, M.: ibid. *90*, 824 (1986)
93. Furlong, D. N., Grieser, F., Hayes, D., Hayes, R., Sasse, W., Wells, D.: J. Phys. Chem. *90*, 2388 (1986)
93a. Sobczynski, A., Bard, A. J., Campion, A., Fox, M. A., Mallouk, T., Webber, S. E., White, J. M.: J. Phys. Chem. *91*, 3316 (1987)
94. Curran, J. S., Lamouche, D.: ibid. *87*, 5405 (1983)
95. Gerischer, H.: ibid. *88*, 6096 (1984)
96. Nosaka, Y., Ishizuka, Y., Miyama, H.: Ber. Bunsenges. Phys. Chem. *90*, 1199 (1986)
97. Dimitrijević, N. M., Li, S., Grätzel, M.: J. Am. Chem. Soc. *106*, 6565 (1984)
98. Taqui Khan, M. M., Bhardwaj, R. C., Jadhav, C. M.: J. Chem. Soc. Chem. Commun. 1690 (1985)
99. Kuczynski, J. P., Milosavljevic, B. H., Thomas, J. K.: J. Phys. Chem. *88*, 980 (1984)
99a. Krishnan, M., White, J. R., Fox, M. A., Bard, A. J.: J. Am. Chem. Soc. *105*, 7002 (1983)
100. Mau, A. W.-H., Huang, C.-B., Kakuta, N., Bard, A. J., Campion, A., Fox, M. A., White, J. M., Webber, S. E.: ibid. *106*, 6537 (1984)
101. Meissner, D., Memming, R., Kastening, B.: Chem. Phys. Lett. *96*, 34 (1983)
102. Yanagida, S., Azuma, T., Sakurai, H.: Chem. Lett. (Chem. Soc. Japan) (1982), 1069
103. Bücheler, J., Zeug, N., Kisch, H.: Angew. Chem. *94*, 792 (1982)
104. Reber, J.-F., Meier, K.: J. Phys. Chem. *88*, 5903 (1984)
105. Henglein, A., Gutiérrez, M., Fischer, Ch.-H.: Ber. Bunsenges. Phys. Chem. *88*, 170 (1984)
106. Yanagida, S., Azuma, T., Kawakami, H., Kizumoto, H., Sakurai, H.: J. Chem. Soc., Chem. Commun. (1984), 21
107. Zeug, N., Bücheler, J., Kisch, H.: J. Am. Chem. Soc. *107*, 1459 (1985)
107a. Kisch, H., Fernández, A., Millini, R.: Chem. Ber. *119*, 3473 (1986)
107b. Kisch, H., Schlamann, W.: Chem. Ber. *119*, 3483 (1986)
108. Alfassi, Z., Bahnemann, D., Henglein, A.: J. Phys. Chem. *86*, 4656 (1982)
109. Gurevich, Y. Y., Pleskov, Y. V., Rotenberg, Z. A.: Photoelectrochemistry, Consultants Bureau, New York and London 1980
110. Haase, M., Weller, H., Henglein, A.: J. Phys. Chem. in press
111. Duonghong, D., Ramsden, J., Grätzel, M.: J. Am. Chem. Soc. *104*, 2977 (1982)
112. Kuczynski, J., Thomas, J. K.: Langmuir *1*, 158 (1985)
113. Nosaka, Y., Fox, M. A.: J. Phys. Chem. *90*, 6521 (1986)
114. Serpone, N., Sharma, D. K., Jamieson, M. A., Grätzel, M., Ramsden, J. J.: Chem. Phys. Lett. *115*, 473 (1985)

115. Albery, W. J., Brown, G. T., Darwent, J. R., Saievar-Iranizad, E.: J. Chem. Soc. Faraday Trans. 1, *81*, 1999 (1985)
116. Albery, W. J., Bartlett, P. N., Porter, J. D.: J. Electrochem. Soc. *131*, 2892 (1984)
117. Albery, W. J., Bartlett, P. N., Porter, J. D.: ibid. *131*, 2896 (1984)
118. White, J. R., Bard, A. J.: J. Phys. Chem. *89*, 1947 (1985)
119. Henglein, A., Kumar, A., Janata, E., Weller, H.: Chem. Phys. Lett., *132*, 133 (1986)
120. Jain, R. K.: Opt. Eng. *21*, 199 (1982)
121. Jain, R. K., Lind, R. C.: J. Opt. Soc. Amer. *73*, 647 (1983)
122. Ricard, D., Roussignol, Ph., Flytzanis, C.: Optics Lett. *10*, 511 (1985)
123. Flytzanis, C., Hache, F., Ricard, D., Roussignol, Ph.: in The Physics and Fabrication of Microstructures and Microdevices (eds. Kelly, M. J., Weisbuch, C.): Springer-Verlag Berlin 1986
124. Baral, S., Fojtik, A., Weller, H., Henglein, A.: J. Am. Chem. Soc. *108*, 375 (1986)
124a. Kamat, P. V., Dimitrijević, N. M., Fessenden, R. W.: J. Phys. Chem. *91*, 396 (1987)
125. Baur, E., Perret, A.: Helv. Chim. Acta *7*, 910 (1924)
126. Krasnovsky, A. A., Brin, G. P.: Dokt. Akad. Nauk *147*, 656 (1962)
126a. Bernas, A.: J. Phys. Chem. *68*, 2047 (1964)
127. Fujishima, A., Honda, K.: Nature (London) *237*, 37 (1972)
128. Lehn, J.-M., Sauvage, J.-P., Ziessel, R.: Nouv. J. Chim. *4*, 623 (1980)
129. Lehn, J.-M., Sauvage, J.-P., Ziessel, R., Hilaire, L.: Israel J. Chem. *22*, 168 (1982)
130. Franck, S. N., Bard, A. J.: J. Phys. Chem. *81*, 1484 (1977)
131. Kraeutler, B., Bard, A. J.: J. Am. Chem. Soc. *99*, 7729 (1977); Nouv. J. Chim. *3*, 31 (1979)
132. Bard, A. J.: J. Photochem. *10*, 50 (1979)
133. Reiche, H., Bard, A. J.: J. Amer. Chem. Soc. *101*, 3127 (1979)
134. Bard, A. J.: Science *207*, 139 (1980)
135. Bard, A. J.: J. Phys. Chem. *86*, 172 (1982)
136. Fox, M. A.: Acc. Chem. Res. *16*, 314 (1983)
137. Pichat, P.: in Organic Phototransformations in Non Homogeneous Media (ed. Fox, M. A.) ACS Symposium Series 278, 21 (1985)
138. Kiwi, J., Borgarello, E., Pelizzetti, E., Visca, M., Grätzel, M.: Angew. Chem. *92*, 663 (1980)
139. Grätzel, M.: Acc. Chem. Res. *14*, 376 (1981)
140. Grätzel, M., Kalyanasundaram, K., Kiwi, J. in Solar Energy Materials, Structure and Bonding, Vol. 49, Springer Verlag, Berlin, Heidelberg, New York, 1982, 37
141. Bahnemann, D., Henglein, A., Lilie, J., Spanhel, L.: J. Phys. Chem. *88*, 709 (1984)
142. Bahnemann, D., Henglein, A., Spanhel, L.: Faraday Disc. Chem. Soc. *78*, 151 (1984)
143. Moser, J., Grätzel, M.: Helv. Chim. Acta *65*, 1436 (1982)
144. Moser, J., Grätzel, M.: J. Am. Chem. Soc. *105*, 6547 (1983)
144a. Spanhel, L., Weller, H., Henglein, A.; J. Amer. Chem. Soc. in press
145. Howe, R. F., Grätzel, M.: J. Phys. Chem. *89*, 4495 (1985)
146. Rothenberger, G., Moser, J., Grätzel, M., Serpone, N., Sharma, D. K.: J. Am. Chem. Soc. *107*, 8054 (1985)
147. Grätzel, M., Frank, A. J.: J. Phys. Chem. *86*, 2964 (1982)
148. Grätzel, M., Moser, J.: Proc. Natl. Acad. Sci. *80*, 3129 (1983)
149. Dimitrijević, N. M., Savić, D., Mićić, O. I., Nozik, A. J.: J. Phys. Chem. *88*, 4278 (1984)
150. Brown, G. T., Darwent, J. R., Fletcher, P. D. I.: J. Am. Chem. Soc. *107*, 6446 (1985)
150a. Darwent, J. J. Chem. Soc. Faraday Trans. II, *82*, 2323 (1986)
151. Albery, W. J., Bartlett, P. N., Wilde, C. P., Darwent, J. R.: ibid. *107*, 1854 (1985)
152. Brown, G. T., Darwent, J. R.: J. Chem. Soc. Chem. Commun. (1985) 98
153. Reichman, B., Byvik, C. E.: J. Phys. Chem. *85*, 2255 (1981)
154. Dabestani, R., Wang, X., Bard, A. J., Campion, A., Fox, M. A., Webber, S. E., White, J. M.: ibid. *90*, 2729 (1986)
155. Chandrasekaran, K., Thomas, J. K.: Chem. Phys. Lett. *99*, 7 (1983)
156. Kamat, P. V.: J. Photochem. *28*, 513 (1985)
157. Kamat, P. V.: J. Chem. Soc. Faraday Trans. I, *81*, 509 (1985)
158. Kamat, P. V.: Langmuir *1*, 608 (1985)
159. Kölle, U., Moser, J., Grätzel, M.: Inorg. Chem. *24*, 2253 (1985)
160. Nakahira, T., Grätzel, M.: J. Phys. Chem. *88*, 4006 (1984)
161. Houlding, V., Geiger, T., Kölle, U., Grätzel, M.: J. Chem. Soc. Chem. Comm. 681 (1982)

162. Warman, J. M., de Haas, M. P., Grätzel, M., Infelta, P. P.: Nature *310*, 305 (1984)
163. Rossetti, R., Beck, S. M., Brus, L. E.: J. Am. Chem. Soc. *104*, 7322 (1982)
164. Rossetti, R., Beck, S. M., Brus, L. E.: ibid. *106*, 980 (1984)
165. Kamat, P. V., Fox, M. A.: Chem. Phys. Lett. *102*, 379 (1983)
166. Moser, J., Grätzel, M.: J. Am. Chem. Soc. *106*, 6557 (1984)
166a. Moser, J., Grätzel, M., Sharma, D. K., Sperpone, N.: Helv. Chim. Acta *68*, 1686 (1985)
167. Fessenden, R. W., Kamat, P. V.: Chem. Phys. Lett. *123*, 233 (1986)
168. Kamat, P. V., Chauvet, J.-P., Fessenden, R. W.: J. Phys. Chem. *90*, 1389 (1986)
169. Duonghong, D., Serpone, N., Grätzel, M.: Helv. Chim. Acta *67*, 1012 (1984)
170. Rossetti, R., Brus, L. E.: J. Amer. Chem. Soc. *106*, 4336 (1984)
171. Mills, A., Porter, G.: J. Chem. Soc. Faraday Trans. I, *78*, 3659 (1982)
172. Sakata, T., Kawai, T.: Chem. Phys. Lett. *80*, 341 (1981)
173. Sakata, T., Kawai, T., Hashimoto, K.: ibid. *88*, 50 (1982)
174. Hashimoto, K., Kawai, T., Sakata, T.: Nouv. J. Chim. *7*, 249 (1983)
175. Baba, R., Nakabayashi, S., Fujishima, A., Honda, K.: J. Phys. Chem. *89*, 1902 (1985)
176. Prahov, L. T., Disdier, J., Herrmann, J.-M., Pichat, P.: Int. J. Hydrogen Energy *9*, 397 (1984)
177. Sato, S., White, J. M.: Chem. Phys. Lett. *72*, 83 (1980)
178. Harriman, A., Porter, G., Walters, P.: J. Chem. Soc., Faraday Trans. *77*, 2373 (1981)
179. Mills, A.: J. Chem. Soc. Dalton Trans. (1982) 1213
180. Blondeel, G., Harriman, A., Williams, D.: Solar Energy Materials *9*, 217 (1983)
181. Blondeel, G., Harriman, A., Porter, G., Urwin, D., Kiwi, J.: J. Phys. Chem. *87*, 2629 (1983)
182. Mau, A. W.-H., Overbeek, J. M., Loder, J. W., Sasse, W. H. F.: J. Chem. Soc., Faraday Trans. 2, *82*, 869 (1986)
183. Bahnemann, D., Fischer, Ch.-H., Janata, E., Henglein, A.: J. Chem. Soc., Faraday Trans. I, *83*, 2559 (1987)
184. Muraki, H., Saji, T., Fujihira, M., Aoyagui, S.: J. Electroanal. Chem. *169*, 319 (1984)
185. Duonghong, D., Grätzel, M.: J. Chem. Soc. Chem. Comm. 1597 (1984)
186. Kiwi, J., Morrison, C.: J. Phys. Chem. *88*, 6146 (1984)
187. Kiwi, J., Grätzel, M.: ibid. *88*, 1302 (1984)
188. Boe Gu, Kiwi, J., Grätzel, M.: Nouv. J. Chim. *9*, 539 (1985)
189. Boonstra, A. H., Mutsaers, C. A. H. A.: J. Phys. Chem. *79*, 2025 (1975); *79*, 1940 (1975); *79*, 1694 (1975)
190. Harbour, J. R., Tromp, J., Hair, M. L.: Can. J. Chem. *63*, 204 (1985)
191. Buxton, G. V., Rhodes, T., Sellers, R. M.: Nature (London) *295*, 583 (1982)
192. Buxton, G. V., Rhodes, T., Sellers, R. M.: J. Chem. Soc., Faraday Trans. 1, *79*, 2961 (1983)
193. Stramel, R. D., Thomas, J. K.: J. Colloid and Interfacial Science, *110*, 121 (1980)
194. Nakanishi, H., Sanchez, C., Hendewerk, M., Somorjai, G. A.: Mat. Res. Bull. *21*, 137 (1986)
195. Darwent, J. R., Mills, A.: J. Chem. Soc. Faraday Trans. 2, *78*, 359 (1982)
196. Erbs, W., Desilvestro, J., Borgarello, E., Grätzel, M.: J. Phys. Chem. *88*, 4001 (1984)
197. Kiwi, J., Grätzel, M.: J. Chem. Soc. Faraday Trans. II, *78*, 931 (1982)
198. Nenadović, M. T., Rajh, T., Mićić, O. I., Nozik, A. J.: J. Phys. Chem. *88*, 5827 (1984)
199. Darwent, J. R.: J. Chem. Soc. Chem. Comm. 798 (1982)
200. Akid, R., Darwent, J. R.: J. Chem. Soc. Dalton Trans. (1985) 395
201. Cunningham, J., Kelly, J. J., Penny, A. L.: J. Phys. Chem. *75*, 617 (1971)
202. Cunningham, J., Penny, A. L.: ibid. *76*, 2353 (1972)
203. Cunningham, J., Zainal, H.: ibid. *76*, 2362 (1972)
204. Morrison, S. R., Freund, T.: J. Chem. Phys. *47*, 1543 (1967)
205. Gerischer, H.: Surface Sci. 97 (1969)
206. Hauffe, K., Rauge, J., Volenik, K.: Ber. Bunsenges. Phys. Chem. *74*, 286 (1970)
207. Hirschwald, W.: Current topics in material science, North-Holland, Amsterdam *6*, 109 (1980); *7*, 143 (1981)
208. Koch, U., Fojtik, A., Weller, H., Henglein, A.: Chem. Phys. Lett. *122*, 507 (1985)
209. Bahnemann, D., Kormann, C., Hoffmann, M. R.: J. Phys. Chem., *91*, 3789 (1987)
209a. Haase, M., Weller, H., Henglein, A.: J. Phys. Chem. in press
209b. Spanhel, L., Weller, H., Henglein, A.: J. Am. Chem. Soc., in press
210. Hada, H., Tanemura, H., Yonezawa, Y.: Bull. Chem. Soc. Japan *51*, 3154 (1978)
211. Tanemura, H., Yonezawa, Y., Hada, H.: Ber. Bunsenges. Phys. Chem. *89*, 837 (1985)

212. Harbour, J. R., Hair, M. L.: J. Phys. Chem. *83*, 652 (1979)
213. Harriman, A.: J. Chem. Soc. Faraday Trans. 1, *79*, 2875 (1983)
214. Fojtik, A., Henglein, A., Katsikas, L., Weller, H.: Chem. Phys. Lett. *138*, 535 (1987)
215. Erbs, W., Kiwi, J., Grätzel, M.: Chem. Phys. Lett. *110*, 648 (1984)
216. Lume-Pereira, C., Baral, S., Henglein, A., Janata, E.: J. Phys. Chem. *89*, 5772 (1985)
217. Baral, S., Lume-Pereira, C., Janata, E., Henglein, A.: ibid. *89*, 5779 (1985)
218. Berry, C. R.: Phys. Rev. *161*, 611 (1967)
219. Meehan, E. J., Miller, J. K.: J. Phys. Chem. *72*, 1523 (1968)
220. Rossetti, R., Ellison, J. L., Gibson, J. M., Brus, L. E.: J. Chem. Phys. *80*, 4464 (1984)
221. Chestnoy, N., Hull, R., Brus, L. E.: ibid. *85*, 2237 (1986)
222. Tsu, R., Esaki, L.: Appl. Phys. Lett. *22*, 562 (1975)
223. Bastard, G., Mendez, E. E., Chang, L. L., Esaki, L.: Phys. Rev. *B 26*, 1974 (1982)
224. Dingle, R.: in Advances in Solid State Physics (ed. Queisser, H. J.) Pergamon Viehweg, Braunschweig *15*, 21 (1975)
225. Shinozuka, Y., Matsuura, M.: Phys. Rev. *B 28*, 4878 (1983)
226. Bimberg, D., Christen, J., Steckenborn, A.: in Advances in Solid State Physics (ed. Queisser, H. J.) Pergamon-Viehweg, Braunschweig *53*, 136 (1984)
227. Nozik, A. J., Thacker, B. R., Turner, J. A., Olson, J. M.: J. Am. Chem. Soc. *107*, 7805 (1985)
228. Laviron, M., Averbuch, P., Godfrin, H., Rapp, R. E.: J. Phys. Lett. *44*, L-1021 (1983)
229. Petroff, P. M., Gossard, A. C., Logan, R. A., Wiegmann, W.: Appl. Phys. Lett. *44*, 635 (1982)
230. Ekimov, A. I., Onushchenko, A. A.: Pisḿa Zh. Eksp. Teor. Fis. *40*, 337 (1984)
230a. Wang, Y., Herron, N. J.: Phys. Chem. *91*, 257 (1987)
231. Tricot, Y.-M., Fendler, J. H.: J. Phys. Chem. *90*, 3369 (1986)
232. Watzke, H. J., Fendler, J. H.: ibid., *91*, 854 (1987)
233. Efros, Al. L., Efros, A. L.: Sov. Phys. Semicond. *16*, 772 (1982)
234. Sandroff, C. J., Kelty, S. P., Hwang, D. M.: J. Chem. Phys. *85*, 5337 (1986); Sandroff, C. J., Hwang, D. M., Chung, W. M.: Phys. Rev. *B 33*, 5953 (1986)
234a. Mićić, O. I., Nenadovic, M. T., Peterson, M. W., Nozik, A. J.: J. Phys. Chem. *91*, 1295 (1987)
235. Brus, L. E.: J. Chem. Phys. *79*, 5566 (1983); *80*, 4403 (1984)
236. Brus, L. E.: J. Phys. Chem. *90*, 2555 (1986)
237. Wolfe, J. P., Mysyrowicz, A.: Scientific American *250*, 3, 70 (1984)
238. Fischer, Ch.-H., Weller, H., Fojtik, A., Lume Pereira, C., Janata, E., Henglein, A.: Ber. Bunsenges. Phys. Chem. *90*, 46 (1986)
239. Rossetti, R., Hull, R., Gibson, J. M., Brus, L. E.: J. Chem. Phys. *82*, 552 (1982)
240. Henglein, A., Fojtik, A., Weller, H.: Ber. Bunsenges. Phys. Chem., *91*, 441 (1987)
241. Nozik, A. J., Williams, F., Nenadović, M. T., Rajh, T., Mićić, O. I.: J. Phys. Chem. *89*, 397 (1985)
242. Rossetti, R., Hull, R., Gibson, J. M., Brus, L. E.: J. Chem. Phys. *83*, 1406 (1985)
243. Nedeljković, J. M., Nenadović, M. T., Mićić, O. I., Nozik, A. J.: J. Phys. Chem. *90*, 12 (1986)
244. Leland, J. K., Bard, A. J.: Chem. Phys. Lett., *139*, 453 (1987)
245. Weller, H., Schmidt, H. M., Koch, U., Fojtik, A., Baral, S., Henglein, A., Kunath, W., Weiss, K., Dieman, E.: Chem. Phys. Lett. *124*, 557 (1986)
246. Schmidt, H. M., Weller, H.: Chem. Phys. Letters *129*, 615 (1986)
247. Kayanuma, Y.: Solid State Comm. *59*, 405 (1986)

Author Index Volumes 101–143

Contents of Vols. 50–100 see Vol. 100
Author and Subject Index Vols. 26–50 see Vol. 50

The volume numbers are printed in italics

Bei Fragen zur Produktsicherheit wenden Sie sich bitte an:
If you have any questions regarding product safety,
please contact:

Walter de Gruyter GmbH
Genthiner Straße 13
10785 Berlin
productsafety@degruyterbrill.com